ALTERNATIVES FOR INSPECTING OUTER CONTINENTAL SHELF OPERATIONS

Committee on Alternatives for Inspection of Outer Continental Shelf Operations

Marine Board

Commission on Engineering and Technical Systems
National Research Council

NATIONAL ACADEMY PRESS
Washington, D.C. 1990

NOTICE: The project that is the subject of this report was approved by the Governing Board of the National Research Council, whose members are drawn from the councils of the National Academy of Sciences, the National Academy of Engineering, and the Institute of Medicine. The members of the panel responsible for the report were chosen for their special competences and with regard for appropriate balance.

This report has been reviewed by a group other than the authors according to procedures approved by a Report Review Committee consisting of members of the National Academy of Sciences, the National Academy of Engineering, and the Institute of Medicine.

The project described in this report is support by Cooperative Agreement No. 14-12-0001-30416 between the Minerals Management Service of the U.S. Department of the Interior and the National Academy of Sciences.

Library of Congress Catalog Card No. 89-64416
International Standard Book Number 0-309-04227-5

Additional copies of this report are available from:

National Academy Press
2101 Constitution Avenue, NW
Washington, D.C. 20418

Printed in the United States of America

COMMITTEE ON ALTERNATIVES FOR INSPECTION OF OUTER CONTINENTAL SHELF OPERATIONS

G. H. PATRICK BURSLEY, *Chairman*, U.S. Coast Guard (retired), Bethesda, Maryland
KENNETH E. ARNOLD, Paragon Engineering Services, Inc., Houston, Texas
LUDWIG BENNER, JR., Events Analysis, Inc., Oakton, Virginia
FLOYD E. GARROT, Hot Springs Village, Arkansas
STAN KAPLAN, Pickard, Lowe and Garrick, Inc., Newport Beach, California
ROBERT P. MEEK, ECOMAR, Inc., Goleta, California
DICK H. PINER, JR., T.E.S.T., Inc., Gretna, Louisiana
MIKLOS D. G. SALAMON, Colorado School of Mines, Golden, Colorado
EUGENE C. SMITH, Eugene C. Smith, Inc., New Orleans, Louisiana
DEMO A. STAVROS, Eastern Michigan University, Ypsilanti, Michigan
EDWIN L. ZEBROSKI, APTEC Engineering Service, Inc., Sunnyvale, California

STAFF

DONALD W. PERKINS, Associate Director
COURTLAND S. LEWIS, Editor
CARLA D. MOORE, Project Assistant

MARINE BOARD

STAFF

PREFACE

The safe operation of offshore oil and gas development and production facilities on the outer continental shelf (OCS)[1] is important to public acceptance of these operations and the continued ability of the United States to develop what has become the source of one-fourth of the nation's gas and one-eighth of its oil. The Outer Continental Shelf Lands Act of 1978 (OCSLA) requires regular inspections of operations to promote safe practices to protect personnel, facilities, and the environment; and to conserve offshore resources. The Minerals Management Service (MMS) of the Department of the Interior has the responsibility for prescribing safe practices and inspecting oil and gas operations on the OCS. Nevertheless, the primary responsibility for establishing a safe working environment and for conducting the necessary inspections to achieve safe practices to carry out the congressional safety mandate is with the oil and gas leaseholders of the offshore facilities.

In recent years, there have been few major accidents on the OCS involving exploration and production operations under MMS jurisdiction. Even so, recent disasters in the North Sea, and accidents involving spills from pipelines adjacent to platforms and from vessels (which are not in MMS's purview), emphasize the need to have an aggressive safety inspection program and to continually seek ways to enhance its effectiveness. With these needs in mind, the MMS requested that the Marine Board investigate alternative strategies for the inspection of OCS operations and that it recommend improvements in operational safety and environmental protection inspection practices.

Previous Marine Board studies–especially *Implementing Best Available and Safest Technologies for Offshore Oil and Gas* (1979), *Safety and Offshore Oil* (1981), and *Safety Information and Management on the OCS* (1984)–stress the need for an evolutionary OCS regulatory regime and suggest that MMS try alternative inspection, enforcement, and compliance approaches. This report is intended to assist MMS to find ways to benefit from its experience with its OCS inspection program as well as that of the Coast Guard, which has an extensive program of marine safety inspections affecting OCS operations.

The report assesses alternatives for revising the present MMS inspection system. The assessment was conducted by the Committee on Alternatives for Inspection of Outer Continental Shelf Operations, operating under the auspices of the Marine Board of the National Research Council.

The committee was charged with developing inspection strategies to improve safety and the effectiveness of the inspection process. The methodology stipulated for conducting the study was to

---

[1]The OCS is that portion of the submerged continental margin that is subject to U.S. jurisdiction. For the purpose of this report, the OCS extends from a state's offshore boundary out to the limit of economic exploitation.

- review the current OCS inspection program;
- appraise inspection practices elsewhere for lessons learned, including inspection of OCS operations not under direct federal jurisdiction (i.e., those in state waters) as well as analogous inspection programs in other industries;
- develop alternatives for conducting inspection programs and assess their advantages and disadvantages; and
- recommend alternative inspection procedures that may be more effective and efficient than present ones and that may make industry more conscious of its accountability for safety.

Committee members were selected for their expertise and to achieve a balanced viewpoint. The members have experience in the fields of OCS oil and gas operations, safety management, safety systems analysis, safety in the nuclear power and mining industries, and public interest in environmental protection. The principle guiding the makeup of the committee and its work, consistent with the policy of the National Research Council, was not to exclude the bias that might accompany expertise vital to the study, but to seek balance and fair treatment.

MMS inspection programs and data bases were reviewed to determine how present inspections are conducted, the extent of the MMS overview on OCS platforms, the inspection functions undertaken by industry personnel, and the OCS safety record. The committee visited OCS operations in the Gulf of Mexico and off the coast of Southern California, and met district and regional MMS managers. In support of this study, the Coast Guard provided information and data concerning safety statistics and its approach to OCS inspection. Information regarding state and foreign inspection practices was obtained through the personal contacts of individual committee members.

This report describes the various inspection alternatives the committee examined; presents the evaluation criteria used; and evaluates each alternative against the criteria. In evaluating alternatives, the committee did not limit itself to current inspection functions. It also examined functions that could and should be included in the safety inspection program to enhance safety in the workplace. Chapter 7 of the report focuses on these additional inspection activities at length. The committee's findings, conclusions, and recommendations are set out in the final chapter.

CONTENTS

EXECUTIVE SUMMARY

About half the world's offshore oil and gas platforms are found in U.S. waters–about one-fourth of U.S. gas production and one-eighth of domestic oil production comes from operations on the outer continental shelf (OCS). On the whole–particularly in recent years–there have been few major accidents involving exploration and production operations under the jurisdiction of the Minerals Management Service (MMS) of the Department of the Interior. However, the massive oil spill from the *Exxon Valdez* on Alaska's Prince William Sound in March 1989 demonstrated dramatically the serious environmental consequences of a major accident in the oil and gas industry. That spill did not involve offshore oil exploration or production, but media focus on it overshadowed a pipeline explosion under Department of Transportation jurisdiction that had taken place aboard a Gulf of Mexico oil production platform only a few days earlier. That explosion killed 7 workers and injured 10 others–a reminder that vigilance in OCS operations cannot be relaxed.

The people of the United States have a right to expect that oil and gas drilling and production operations on the OCS will be conducted with appropriate regard for safety of the public, operating personnel, and the environment. This expectation is explicit and implicit in public legislation and policy governing offshore operations. It is likewise embodied in the leases granted to offshore operators. By law, leases are to be granted only to competent operators who can demonstrate that they have the requisite resources to carry out operations in a safe and environmentally sound manner. The Minerals Management Service (MMS) of the U.S. Department of the Interior is charged with ensuring that only qualified operators are granted leases and that they carry out operations in a safe and environmentally sound manner.

By law the primary responsibility for safety is placed on the operator. The MMS is charged with ensuring that the operator carries out this responsibility in a fully satisfactory manner. The scope of MMS's responsibility extends to the comprehensive regulation of fixed platforms and drilling and production operations on both fixed and mobile platforms. To carry out its regulatory responsibility, the MMS has promulgated regulations regarding safe operation, including the installation, maintenance, and periodic testing of various safety devices. The MMS also has an inspection program to verify that operators comply with regulations and operate in a safe and environmentally sound manner. The inspection program is centered around compliance with a listing of "potential incidents of non-compliance," or PINCs. Most PINCs relate to the verification of set-points and functioning of specific safety devices, such as pressure gauges and pressure relief valves. Only a few PINCs address general safety issues: for example, "are operations performed in a safe and workmanlike manner?" and "are necessary precautions taken?"

The Outer Continental Shelf Lands Act of 1978 (OCSLA) specifies that all OCS facilities must have at least one scheduled onsite inspection annually. In addition to the scheduled annual inspection for each platform, the MMS is required to conduct periodic unannounced "spot" inspections (administratively it has set a target of 50 percent of the drilling rigs and 10 percent of the production platforms each year). During these inspections, the MMS inspector works through the items on the PINC list–all items on the list in the case of annual inspections, some items in spot inspections–witnessing tests of the items as they are performed by the operator's personnel. Any deviation found constitutes a violation, and is termed an "incident of non-compliance" (INC), which the operator must correct, subject to sanctions. The sanctions range from a warning, to a shut-in [stoppage] of specific operations or the entire operation, to civil penalties, to criminal penalties. INCs that require a total shut-in are infrequent, and civil penalties are rarely imposed. Many INCs are corrected immediately, or if they involve a faulty piece of equipment, it is isolated, taken out of service, and repaired without a complete shut-in of the entire faculty.

Tests conducted during inspections are identical to the routine and repetitive tests not witnessed by MMS but required periodically as a matter of safe working practice. Actual tests witnessed by MMS inspectors under the OCSLA comprise less than 10 percent of all those MMS requires (Table 2-1).

Despite OCSLA requirements, the MMS currently does not conduct an annual scheduled inspection on all platforms, nor does it conduct as many spot inspections as it would like due to limited inspection resources (dollars and personnel). Also, during the annual inspections not every PINC item is inspected–although it is required by regulation. The growing complexity of production platforms and the greater distances of new facilities from shore in the Gulf of Mexico are making it increasingly difficult for MMS to satisfy this requirement of the law. In addition, government work rules relating to overtime, time spent offshore, length of workday, and other factors, and the necessity for helicopter transportation to platforms (with helicopters and pilots frequently standing by for most of the day while the inspector does his work), have resulted in inspections that are expensive and inefficient.

The MMS therefore asked the committee to explore alternative ways of deploying MMS inspection resources to carry out its safety responsibilities. In particular, MMS seeks an alternative to annual inspection of all safety devices on the platforms so that it can focus its resources more intensively on particular OCS facilities and operations that need them. In brief, MMS seeks to gain greater efficiency in the use of inspection resources and a greater assurance that it is carrying out its mission as effectively as possible.

The committee reviewed the situation in detail and supports the MMS position that it can be more effective if given the freedom to apply its resources selectively to the areas of greatest need.[1] In arriving at this conclusion, the committee evaluated the current inspection program as well as five possible alternative inspection programs:

1. increased inspection onsite by MMS;
2. inspection of a sampling of PINCs during annual inspections, and increased spot inspections;
3. annual inspection of a sampling of facilities, and increased spot inspections;
4. third-party inspection with government audit; and
5. self-inspection.

---

[1]An important qualification to this endorsement is that it applies only to production facilities and operations. For drilling facilities and operations, the current program of frequent and comprehensive inspections should continue because of (1) the high frequency of "events" (i.e., accidents) per facility for drilling facilities as compared to production facilities, and (2) the large population of workers on a facility while it is engaged in drilling operations. Additionally, the committee believes that at an early date MMS should introduce the enhanced data collection and safety analysis program discussed in Chapter 7 to the fullest extent possible.

The first three alternatives are modified versions of the present MMS inspection system. The other two involve a fundamental departure from current practice.

The committee evaluated the present program and the five alternatives in view of considerations deemed important from the standpoint of effectiveness and acceptability of any OCS inspection program. They included the following:

- Does the alternative being considered promote safety awareness?
- Does it help to maintain public confidence in the safety of OCS operations?
- Does it use inspection resources efficiently?
- What is the impact on the qualifications and training of the inspector force?
- Does the program provide for identification of safety trends and warnings?
- Does it promote safety performance accountability?
- Is it adaptable to changing circumstances?
- Are there valid precedents?

Based on its assessment of present and alternative inspection programs, the committee concluded that the present program is conducted conscientiously and supports the need to ensure compliance with inspection requirements as stated in the OCSLA. However, the program does not incorporate the necessary data and information collection and analytical activity for systematically upgrading safety performance requirements for OCS operations.

The committee recommended that the current inspection program be modified to enhance its effectiveness at present levels of personnel and funding through implementation of alternative 2 (inspection of a sampling of PINCs–rather than all the PINCs–during the annual inspection, together with increased spot inspections). Such a program would be more efficient and as effective as the present program in ensuring compliance with the PINC list. Importantly, it also would free resources for complementary activities that would significantly improve the present program. The inspector resources made available by having to witness fewer tests should be redirected toward

- increased spot inspections, instituted systematically on the basis of inspection results, operator safety histories, and interviews with the operator personnel; and
- analysis of data to identify emerging safety problems and general safety trends.

The alternative program would continue to require each operator to perform and record in a prescribed format all the scheduled inspections/tests (both those specified by MMS and those in the operator's own facility inspection program) with only selective inspections/tests–derived from a sampling plan–being witnessed by MMS to verify the operator's performance of the inspections/tests. Ultimately, as experience and confidence in the effectiveness of the sampling plans is developed, the inspection program might be modified further into alternative 3, sampling of facilities for annual inspection (with increased spot inspections), in which only a limited number of facilities would be subjected to an intensive annual inspection and one spot inspection per year (as a minimum) would be conducted at the others.

Beyond its principal recommendation, the committee also made a number of observations and recommendations to MMS regarding ways to improve its oversight of safety performance on the OCS. Among these are the following:

The committee recommends that MMS improve its collection, analysis, and use of safety-related data regarding offshore operations. This recommendation is based on the fact that improvements in safety performance derive in large part from past lessons. To learn from past experience demands a clear perception of what that experience has been. This means that the data base on OCS operations must be made more comprehensive and accessible. It must include more information, which must be organized and processed so that proper inferences can be made, conclusions clearly drawn, and lessons readily learned.

The main near-term use of an expanded and enhanced data base would be the development of a sampling plan for selecting PINCs to be inspected. To that end, it would be crucial to have available data on past INCs–their numbers, types, and frequency of occurrence on different kinds of facilities and on specific facilities. Access to INC-to-PINC ratios for each PINC over time would provide a valuable index for determining which PINCs should be sampled, how often, and on which facilities. The safety analysis program should include monitoring and review of operator records ashore and analysis of data and subjective observations to reveal safety trends.

By comparing accident events data (1982 was selected as a recent sample year) with the PINC list, the committee was persuaded that most of the items on the list are reliable safety devices and their failure is not relevant to the kinds of accidents that actually occur. For that reason, the committee believes that items on the PINC list should be continually reviewed, expanded, pruned, or otherwise revised so that the PINC list focuses on the causes of accidents that actually are experienced and their early warning signals. Analysis based on an improved data base would permit this to be done with confidence.

The potential implementation of the selected-facilities inspection alternative (alternative 3) at a later date would require even more extensive data, much of it facility specific. Putting this alternative in place probably would require quantitative indices that characterize and measure the safety of individual offshore operations. Among the factors that should be taken into account in developing sampling indices would be the following:

- the occurrence of safety-related "events" aboard the facility;
- the occurrence of "near misses" (i.e., operational disruptions that did not result in a reportable accident);
- the record of tests and inspections, in terms of safety equipment not working;
- evidence of slipshod operation (e.g., poor maintenance, poor housekeeping, poor recordkeeping);
- the facility design, including features such as location and age;
- evidence of lax safety attitudes of managers, supervisors, or operating personnel;
- the overall safety record of the operator as to all his facilities; and
- the overall safety of all operators in the region or district.

From such quantitative, facility-specific information, a safety rating could be developed for each platform, which would be updated continually with new data. The data base would be kept up to date by requiring that all "event" reports and specified operator's inspection and test results be sent to MMS. Onshore review of records could then comprise a substantial part of the inspection. Actual onsite inspections would be made more efficient by prior analysis of the information in the data base. Inspection and test results submitted by the operator could be checked for consistency with actual field inspection results.

The committee also recommended that MMS encourage its inspectors to look for emerging and changing safety risks on OCS facilities. The position description, job assignments, and reward structure for MMS inspectors should be amended to reflect the importance of risk identification and reporting. Information on mishaps should be disseminated in a manner similar to safety alerts, and should result in appropriate changes in permit requirements, training, and regulations.

The safety of a platform, or any other facility, is not determined just by the quality of its operating manuals and the reliability of its equipment. Major factors are management's safety policy, and the training and attitudes of personnel who manage and operate the facility. MMS should make explicit in its safety management and inspection philosophy that monitoring of safety attitudes of the operators and resulting necessary corrective action are essential. Subjective judgments will be involved, but this fact should not be a deterrent. In fact, MMS inspectors and supervisors should be trained in the techniques for and the importance of monitoring safety attitudes. Moreover MMS should develop programs to provide motivation for operators to "think safety" that are broader and more publicized than the Safety Award for Excellence recognition program currently in place.

The record of safety on the OCS has been good. In terms of injuries and fatalities, OCS drilling and production operations are comparable to other hazardous activities onshore, such as mining and construction. In terms of environmental impact, oil pollution from offshore operations contributes less than any other significant source to the release of hydrocarbons into the marine environment. U.S. offshore industry spillage contributes less than 5 percent of world spillage, and over the past several years the average spill volumes and the amount spilled compared to total production has been reduced.

Thus, MMS and the offshore industry are not faced with the problem of correcting a manifestly poor safety record. The United States has succeeded under its present inspection program in averting the kinds of catastrophic disasters that have befallen the offshore operations of many other nations. Although the evidence of a direct connection is lacking, certainly the activities and vigilance of the federal government have been a factor. However, an increase in the margin of safety on the OCS can be achieved by improving the link between the MMS inspection program and safety performance of the industry. The committee's recommendations are intended to accomplish that end.

A final point made by the committee–and it is a crucial one–relates to attitudes. In enterprises that are subject to inspection by government or other authorities, the operators of the enterprise often gradually drift to the point of view that the responsibility for safety lies with the government and its inspectors. An attitude develops that the operator's responsibility and objective is simply to pass the inspection, an attitude the committee refers to as a "compliance mentality." It is especially likely to develop when inspections are based on a routine checklist approach.

The committee emphasizes its belief that *compliance does not equal safety*. Thus, although it is certainly desirable to have checklists to guide the inspectors, it is important for MMS to ensure that operators do not sink into a compliance mentality. To reiterate: in practice and by law, the operators bear the primary responsibility for safety. The MMS, for its part, is responsible for using the best and most effective means it can devise to motivate operators to meet that responsibility.

1

INTRODUCTION

The federal government has a multifaceted role with respect to activities on the outer continental shelf (OCS). It acts to manage OCS resources, to require that industry operations are conducted safely, to compile and disseminate information, and to foster the development and application of technology that will improve the safety of OCS operations.

Several agencies carry out this role. They include the Minerals Management Service (MMS) of the Department of the Interior, the U.S. Coast Guard (USCG), the Environmental Protection Agency (EPA), and elements of the Department of Transportation and the U.S. Army Corps of Engineers (COE). Inspection responsibility is shared by MMS and USCG. However, it is MMS that is responsible for inspecting operational aspects of OCS activity. Therefore, this report will deal almost exclusively with MMS programs. For a more detailed discussion of the history and purposes of federal management of OCS activities, see Appendix A. Appendix B describes certain aspects of the USCG safety mission on the OCS and recent changes in its inspection strategy.

OVERVIEW OF MMS FUNCTIONS

The Safety Mission

The MMS, in its eighth year of operations in 1989, is charged with promulgating and enforcing regulations over leasing and operations on the OCS, including the safety of life, property, and the environment. Prior to the formation of the MMS, these functions were performed by the U.S. Geological Survey (USGS). The MMS activity is directed at carrying out national policy as specified in the OCS Lands Act (OCSLA), which states in part

> Operations in the outer continental shelf should be conducted in a safe manner by well-trained personnel using technology, precautions, and techniques sufficient to prevent or minimize the likelihood of blowouts, loss of well control, fires, spillages, physical obstruction to other users of the waters or subsoil and seabed, or other occurrences which may cause damage to the environment or to property, or endanger life or health. (43 U.S.C. 1332)

Other pertinent excerpts from the OCSLA appear in Appendix C.

To carry out its safety role, MMS is concerned with most aspects of offshore operations. For example, it has promulgated regulations governing the application of technology to offshore oil and gas operations. As specified by law, leaseholders are required to use the "best available and safest technologies" (BAST) that are economically feasible in all new drilling and production operations, and wherever practical in existing operations.[1]

Leaseholders are required to submit detailed plans governing exploration, development, production, and response to pollution accidents to MMS. Leaseholder preparations for well drilling and production, and their fitness to carry out these operations, are evaluated and verified. MMS maintains files on "events" (fatalities, fires, explosions, blowouts, and injuries associated with these incidents) and other unusual occurrences that halt operations, and inspects leaseholder facilities and operations, applying operating sanctions and civil and criminal penalties for infractions.

MMS inspections are expected to contribute significantly to "assuring safe operations on the OCS as well as assuring that environmental concerns are protected" (U.S. Department of the Interior, 1987c). This concept is significant, in that it conveys a clear intention that the inspection element of the MMS's mission do more than merely ensure leaseholder compliance with the various regulations and requirements. In the final analysis, the effectiveness of the MMS's inspection program must be judged primarily by the degree to which it promotes operational safety and minimizes pollution.

Inspection Budget

The MMS carries out its inspection responsibilities under a major program, the OCS Lands Program. Inspection is part of a subsidiary Regulatory Program under the OCS Lands Program. Approximately $142 million in direct program funds were included in the final 1988 MMS budget. Out of this total, approximately $9 million is allocated for inspections.

OVERVIEW OF THE OCS OPERATIONS

This overview of offshore oil and gas drilling and production operations on the U.S. outer continental shelf describes the types of platforms, the drilling and production facilities in use; their age and geographic distribution; the systems and technologies involved; and the safety and environmental protection considerations that must be addressed.

Facilities

Offshore drilling and production processes and equipment are basically the same as those on land, except that offshore operations require some form of support to protect the facility from water and wave action. This is accomplished by using various types of temporarily positioned vessel-like mobile units, either bottom-supported or floating, and various types of derrick-like structures that are permanently positioned on pilings driven into the seabed (referred to collectively as "offshore platforms"). In the Arctic, ice and gravel islands, concrete islands, and steel-sided caissons are also used to protect facilities from shifting ice as well as water and wave action.

---

[1] In a 1979 report, the Marine Board construed this concept to require "the application of technology in the form of equipment, systems, procedures, and trained workers to ensure the highest degree of operating safety and reliability within reasonable economic limits" (National Research Council, 1979).

Offshore Oil and Gas Operations

Offshore oil and gas operations involve a number of distinct phases: exploration, development, production, and processing.

Initial or exploratory drilling in offshore areas usually is done with drilling equipment carried on some type of temporarily emplaced mobile platform. These mobile platforms, together with the drilling equipment, often are referred to simply as "mobile offshore drilling units," or "MODUs." These mobile platforms include self-elevating jack-ups, submersibles, semisubmersibles, and drillships. Drillships and self-propelled units are capable of changing drilling sites without tug assistance. A drillship is essentially a conventional ship, outfitted with drilling equipment, which is used to drill exploratory wells in deep water far from land. Drillships and semisubmersibles usually are anchored on a site, but they may be dynamically positioned.

If economically recoverable quantities of oil or gas are discovered after exploratory wells are drilled using mobile offshore drilling units, development wells are drilled. (Most deepwater exploratory wells are drilled as expendable wells and either are abandoned, even when productive, or temporarily abandoned and recompleted.) Development drilling can be done by a mobile offshore drilling unit, or by using drilling equipment hoisted onto the permanent production platform by barge-mounted cranes if the offshore platform (usually built in sections onshore) has been erected at the offshore development site.

After the wells are drilled, the drilling equipment is removed, onshore-fabricated production equipment is placed on the platform, and the wells are connected. The production phase of operations then begins, continuing until the wells are depleted. Production commonly involves initial processing (consisting primarily of separation of gas, oil, water, and solids) performed on the platform. The gas and the oil are then transported to shore (generally through pipelines) for additional processing and/or refining.

Offshore platforms are large structures designed to resist hurricane waves and winds. They are becoming even larger and more complex over time as oil and gas discoveries are being made in deeper waters. Both the number of wells per platform and the amount of onboard processing are a function of the economic benefits of distributing the higher costs of deeper water platforms among more wells.[2] An additional influence on the number of wells per platform on the Pacific coast, where existing platforms are very close to shore, is the public pressure to reduce the visual impact of offshore activity. It is estimated that approximately 50 percent of platforms in the Gulf of Mexico contain some processing systems; all Pacific platforms include processing facilities.

Achieving maximum production rates per well and the resort to multiple wells required by increasing offshore costs necessitate large production facilities. Remote locations and the economic need to maximize production require that operating personnel be quartered on the offshore platforms. Drilling operations are conducted around the clock and also require crews to be quartered on the platforms. Multiple-well drilling operations, multiple-well production facilities, high rates of production, and the presence of personnel quarters on platforms, particularly at remote locations, all present additional potential safety hazards to personnel, property, and the environment.

Crew Size

Both drilling and production crews usually spend seven days working offshore and seven days off duty onshore. While at an offshore location, most crews work on two shifts around the clock–12 hours on and 12 hours off. On a drilling facility the personnel complement consists of a total of about 20 to 25 people on average per shift. Three to four of the persons on each shift are salaried

---

[2]Over the period 1978-1987, the Gulf of Mexico average has been 6 wells per platform, and the Pacific average has varied from 44 to 56.

personnel (tool-pushers and supervisors); the remainder are hourly workers. Also on a drilling facility are a number of service personnel who provide specialist functions, such as cementing, logging, major maintenance, and galley and housekeeping support. Some of these reside on the platform, while others are aboard only periodically. Overall the average number of personnel on a drilling facility is 45.

On a production platform, the crew varies with the number of wells and the complexity of the equipment. The majority of the members of a crew work during the day with only a skeleton crew on duty at night. In the Gulf of Mexico, many platforms are unmanned and are serviced from a central platform manned by 20 or fewer people, of whom 5 to 8 perform technical and administrative functions. In the western Gulf, where gas fields are more widely scattered and platforms are smaller, crew sizes tend to be smaller (2 to 10). The average crew size is larger in the Pacific region because these platforms tend to have more oil wells per platform, are larger, and more equipment is required to process the heavy crude oil. The overall average number of on-site personnel on manned production facilities in Gulf of Mexico and Pacific production operations is estimated to be 12.

Geographic Distribution

MMS management of offshore oil and gas resources has been assigned to four OCS regions: Gulf of Mexico, Pacific, Alaska, and Atlantic. Each of these regions is subdivided into a number of areas, as shown in Figures 1-1 through 1-3. Since federal leasing began in 1954, over 537 million acres have been offered for lease and about 45 million acres have been leased (U.S. Department of the Interior, 1986b). Total "bonus" payments (for the leases) to the federal government in this same period exceeded $53 billion, and total oil and gas royalties were over $31 billion.

By far the most production in the offshore regions has been in the Gulf of Mexico. Cumulatively, over 95 percent of the oil and over 99 percent of the gas produced on the OCS has come from the Gulf. However, this dominance appears to be lessening. In 1987 for example, 9.3 percent of the offshore crude oil came from the Pacific region, and new fields now coming on line there could increase production totals if objections to issuing permits are resolved (U.S. Department of the Interior, 1986b, 1987b). There is relatively little offshore activity in the Alaska and Atlantic regions. (Current Alaskan offshore oil and gas comes from fields in state waters.)

The number of active drilling units historically has fluctuated roughly in accordance with the price of oil and the demand for petroleum products. Over the 21-month period from January 1985 to September 1986, for example, the number of active drilling units on the OCS went from 285 to 99. According to more recent data made available to the committee, in February 1988 there were 3,583 production platforms and about 200 drilling units operating in the OCS (MMS, personal communication). Figure 1-4 shows historical data on the number of platforms installed and operating in the Gulf of Mexico.

In the Gulf of Mexico, the depletion of shallow-water, nearshore oil and gas pools over time has led OCS operations into more distant locations and greater average water depths. In the Pacific region, little movement to greater depths has occurred since the initiation of the OCS leasing program. Platforms there are in water as shallow as 95 ft and as deep as 842 ft with the majority located at depths of 150 to 300 ft (MMS, internal memorandum 10/28/88). Table 1-1 gives the range of depths of wells drilled in the Gulf of Mexico and Pacific OCS regions.

Improved technology has made it possible to build conventional design production platforms in water over 1,000-ft deep. Two units are planned for installation off the California coast in 1992 at depths of 1,075 and 1,200 ft. New designs such as the compliant tower and tension-leg platform are expected to lead to emplacement of platforms in waters on both coasts up to 2,000- and 3,000-ft deep, respectively (U.S. Department of the Interior, 1986b).

OCS Planning and Operational Areas

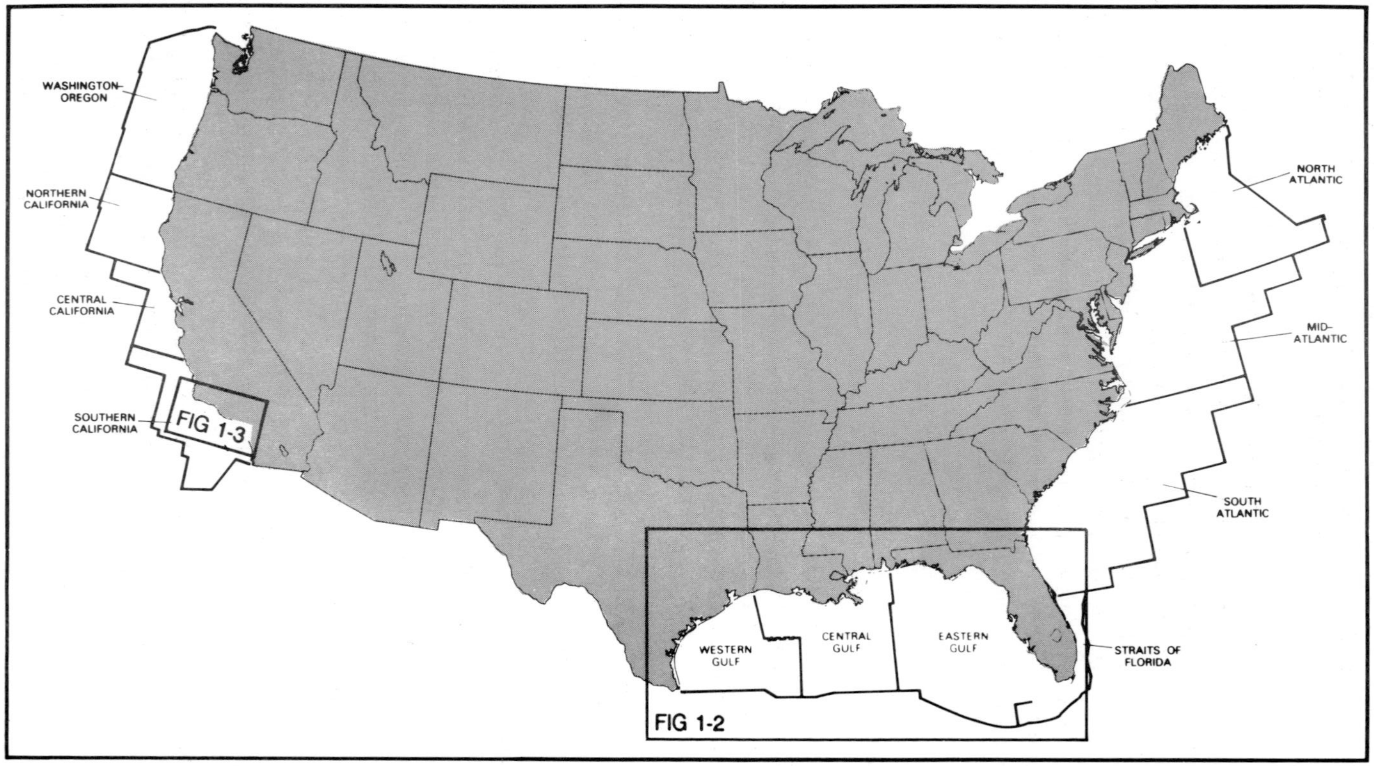

FIGURE 1-1 OCS drilling and production operational areas of the contiguous United States.

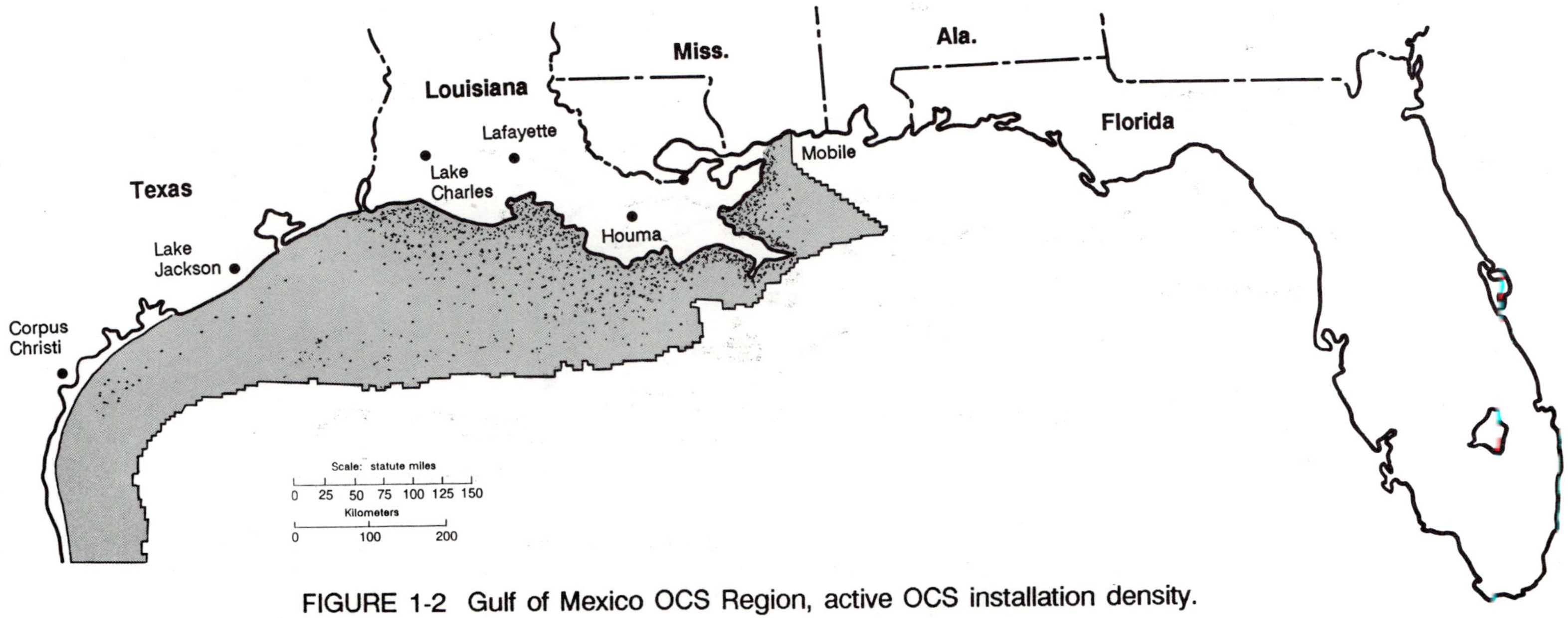

FIGURE 1-2 Gulf of Mexico OCS Region, active OCS installation density.

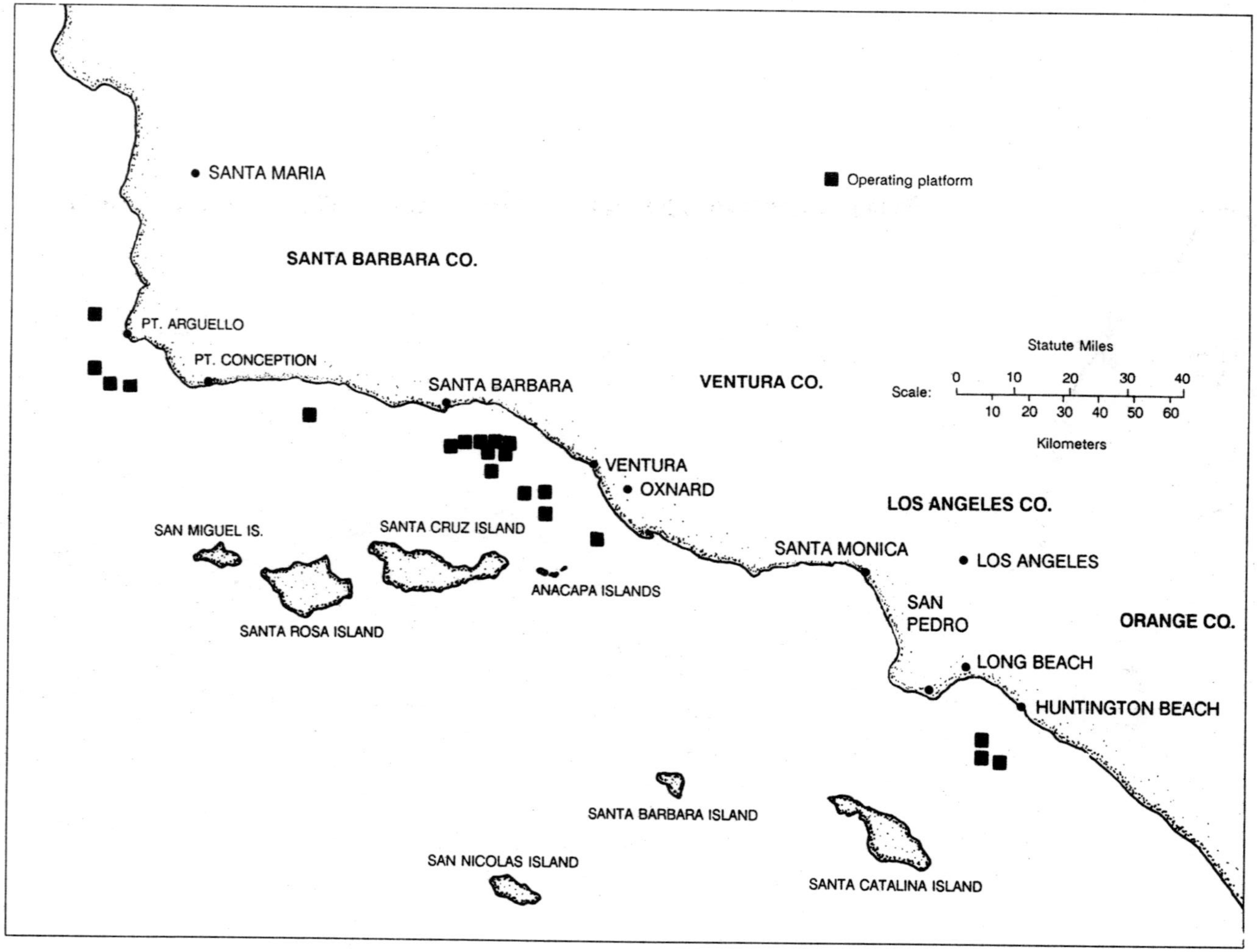

FIGURE 1-3 Pacific OCS Region, active OCS installations.

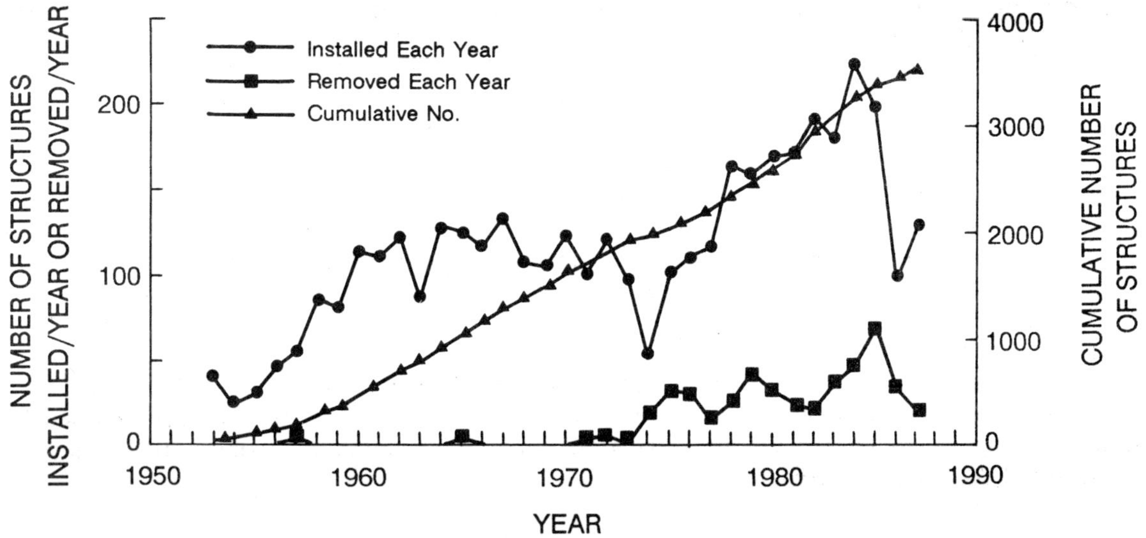

FIGURE 1-4 Gulf of Mexico installations.

TABLE 1-1 Number of Production Structures (OCS waters)[a]

Water Depth (Ft)	Gulf of Mexico	Pacific
0-20	419	0
21-50	1412	0
51-100	830	1
101-150	352	1
151-200	269	6
201-300	246	5
301-400	61	2
401-500	11	1
501-900	7	5
>900	6	0
	3613	21

[a]Number of structures includes nonoperating facilities and therefore does not correspond precisely to the statistics presented in Figure 1-4.

SOURCE: Minerals Management Service, data as of July 1988.

Age Factors

According to MMS data, at the end of 1986, 578 platforms in U.S. waters (roughly 17 percent) were over 25 years old. The average age of production platforms is approximately 15 years. On average, offshore production platforms are designed for and kept in service about 25 years (National Research Council, 1985). This does not mean that platforms older than 25 years are unsound, nor are the systems or components on them necessarily inadequate. However, the age of these units does require that they be subjected to more intensive inspections of their structure and operating systems.

Generally speaking, the service life of an offshore structure is determined by the depletion of economically recoverable oil and gas at the site. Apart from structures damaged by collision, fire, or storm, few offshore structures have been removed because they were no longer structurally sound or serviceable. Platforms are protected from corrosion in the splash zone by coating systems, and under water by cathodic systems. The integrity of the structure more often has been affected by unstable foundation soils and wind and wave forces than by the condition of the structure itself.

The aging of offshore platforms raises concern–not primarily about their structural integrity but about the potential deterioration of onboard systems and facilities. Internal corrosion and erosion of piping on production and processing facilities is an age-related problem that requires sophisticated detection measurements.

Growth in Number of Operating Companies

The MMS has followed a policy of encouraging greater competition at lease sales and has attempted to make bidding more attractive to smaller companies (U.S. Department of the Interior, 1988c). In the August 1988 Gulf of Mexico sale, nine major oil companies accounted for only 52 percent of the high bids on tracts, with the other 48 percent being split among 33 different independent oil companies.

As fields get older, there is an incentive for larger companies, with their relatively high overhead, to farm out the fields to smaller companies that can continue to make a profit due to their lower overhead. In addition, a lease holder may farm out a lease after drilling a discouraging exploratory well because further development would be incompatible with its economic objectives, while another–usually smaller–company, which is not burdened with high sunk investment costs, can use lower cost techniques or may have reached a different geological interpretation.

The proliferation of operating companies can be seen in the fact that in 1983 there were only 20 companies operating up to five leases in the Gulf of Mexico OCS. By 1988, that number had grown to over 65.

The safety implications of a proliferation of operators are hard to document. There are no statistics regarding the accident rate of companies that hold only one or two leases, as against companies that hold many leases. While there is no way to evaluate safety-consciousness of small companies, many companies exhibit the following characteristics that affect overall safety risks:

- Typically they have no in-house safety staff, and usually they have minimal engineering and technical staffs to review field work and train field personnel in safety risks.
- They use contract labor and outside consultants and undertake little, if any, onsite operator supervision.
- Many have limited financial assets, which could lead to deferral of costly safety measures.

Systems and Technologies

The complexity of the inspection task is suggested by the following description of the mechanical, hydraulic, and electrical systems utilized in OCS operations.

Drilling "Making hole" requires a variety of equipment common to all drilling operations. They fall into four system categories (Baker, 1979):

1. *Power generating* systems, which usually use diesel internal combustion engines (although gas turbines also are coming into use) to drive generators. In some locations, where platforms are nearshore or near other platforms containing electrical generation equipment, subsea cables are used to deliver power to the platform.
2. *Hoisting* systems, which include draw-works, weight indicators, catheads, crown and traveling blocks, derricks, and masts used for lifting drilling equipment and pipe.
3. *Rotating* systems, which include the entire drill string, from the swivel below the traveling block to the bit at the bottom of the drill hole, as well as the rotary table on the deck surrounding the drill string.
4. *Circulating* systems, which include the drilling fluid ("mud") and various circulating equipment (the mud pump and associated lines and hoses), as well as auxiliary equipment such as shale shakers and desanders. Circulating systems also include well control devices such as blowout preventers, degassers, and pit level indicators.

Hoisting and rotating systems present the greatest potential for operational accidents resulting in death or injury. Circulation system failures can result in loss of well control and blowouts, accompanying fires, and personnel accidents. Moreover, such systems usually operate intermittently during drilling and require close attention to sensors and gauges while the hole is being made.

Production Oil and gas production requires equipment to process the produced fluids, as illustrated in Figure 1-5. All contain pressurized electrical and/or lifting components that may involve safety hazards. These can be divided into systems as follows:

- *Separation* systems effect the initial separation of well fluids into gas, oil, and water and remove the sediment. Depending on the flowing pressure of the wells or pipeline pressure, several separators operating at different pressures may be required.
- *Oil treating* systems remove small amounts of solids and water left in the oil after separation. Oil treating equipment includes large settling tanks ("gunbarrels"), fired pressure vessels ("heater-treaters") and pressure vessels with electrostatic grids ("chem-electric treaters").
- *Water treating* systems take the water from the separation and oil treating systems (which may contain as much as 2,000 mg oil per liter of water/oil mixture), remove the oil, and dispose of the water at sea (if the residual oil is less than the limitations established by the Environmental Protection Agency) or reinject it in disposal wells.
- *Gas compression* systems use reciprocating or centrifugal compressors to compress the gas, recovered from various low pressure production equipment up to the pressure of the offshore pipelines.
- *Gas treating (dehydrating)* systems condition the gas for sales. Most gas sales contracts require that the water vapor in the gas be limited (normally to 7 pounds water per million standard cubic feet of natural gas). This usually is done using triethylene glycol (TEG) in a closed vessel to absorb water vapor from the wet gas.
- *Well testing* systems consisting of separators and/or treaters are used periodically on individual wells to measure their production of oil, gas, and water, so that the total daily production from the platform can be allocated properly to each well.
- *Utility* systems include equipment designed to provide electricity, fuel gas, instrument air or gas, heat, fresh water, sewage treatment and fire fighting systems.
- *Water injection* systems include systems for processing produced water or seawater for injection into the reservoir for waterflooding.

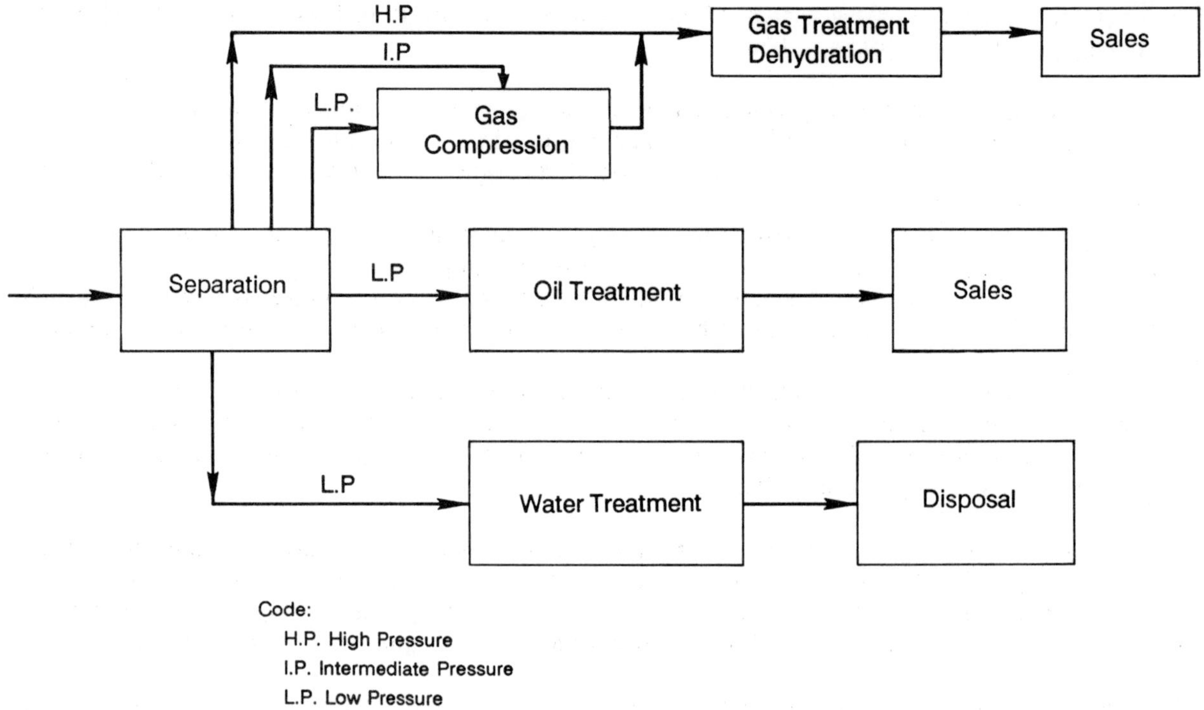

FIGURE 1-5 Typical oil facility block diagram.

These systems can be installed in many combinations. Very few platforms include all systems. For example, a satellite platform for a single well may contain only a simple utility system for instrument gas. Many gas platforms contain a simple utility system, separation system, simple water treating system, and test system.

The control of production systems involves fewer uncertainties than drilling system control. The equipment that is in general use has been proven over many years of application both onshore and offshore. The systems generally operate at a steady-state condition so that monitoring sensors and gauges is relatively easy. Pressure is controlled by sensing the pressure in the gas space of a vessel and regulating the rate at which gas leaves the vessel through an automatic control valve. Level is controlled by sensing the gas/liquid or oil/water level in a vessel and regulating the rate at which liquid leaves through an automatic control valve. Temperature is controlled by sensing process fluid temperature and regulating the flow of either a heat medium or fuel to a burner through an automatic control valve. These three variables can be controlled independently.

System Safety Management

In general, from a production standpoint, the primary concern in the Gulf of Mexico is controlling the natural pressure of the field. In the Pacific, the concern is generating sufficient pressure to extract the oil.

Design Considerations

System safety management in oil and gas processing operations generally involves the extensive use of redundant systems and safety devices as well as generous safety factors in the design of systems. The MMS requires adherence to American Petroleum Institute (API) guidelines (API recommended practice 14C), which provide two levels of protection beyond good process design. Producing wells are controlled by both subsurface and surface safety valves to avoid blowouts and overpressuring downstream processing equipment. Various sensors are used to detect a hazardous situation and automatically close the appropriate valves. All pressure vessels and piping are protected by pressure relieving devices if there is a possibility of exceeding the maximum allowable working pressures. MMS regulations require that all pressure vessels be American Society of Mechanical Engineers (ASME) Code stamped, although noncode vessels may be approved in some instances if hydrotested in accordance with regulations. All piping must be installed in accordance with American National Standards Institute (ANSI) B31.3 pressure piping code, and all electrical equipment must be installed and operated in accordance with the National Electric Code. Various mandated API guidelines and specific MMS requirements are referenced in the *Code of Federal Regulations* (C.F.R.) to amplify and clarify these codes for application to the offshore environment.

Firefighting systems are installed on platforms in accordance with API recommended practices and MMS requirements in work areas and in accordance with Coast Guard requirements for living accommodations. Flame, heat, smoke, and gas detectors generally are placed in potentially high-hazard areas. Many additional safety systems also are installed to achieve shut-in and containment of the various pressurized streams, should a failure occur.

Manufacturing Standards for Safety Equipment

The safety and pollution prevention equipment (SPPE) certification program conducted by MMS requires that all surface safety valves, subsurface safety valves, underwater safety valves, landing nipples, and locks used on the OCS be manufactured under American National Standards Institute/American Society of Mechanical Engineers Standard (ANSI/ASME) SSPE-1, *Quality Assurance and Certification of Safety and Pollution Prevention Equipment used in Offshore Oil and Gas Operations*.

Training

MMS regulations require that all lessee and contract personnel (rotary helpers, derrickmen, drillers, toolpushers, and operator's representatives) be trained in accordance with MMS Standard MMSS-OCS-T-1, *Training and Qualifications of Personnel in Well-Control Equipment and Techniques for Drilling on Offshore Locations* (Second Edition, May 1982). The MMS certifies well-control training programs and conducts onsite evaluations and unannounced audits of those programs to ensure that they are following their approved curricula.

The MMS also requires that all lessee and contract personnel installing, inspecting, testing, and maintaining antipollution safety devices and systems be qualified under a program identified in the American Petroleum Institute's (API) recommended practice API RP T-2, *Qualifications Programs for Offshore Production Personnel Who Work With Antipollution Safety Devices*.

Aspects of Safety on the OCS

Safety in offshore oil and gas operations is of concern from the standpoint of crew safety, the safety of the facility itself, and the protection of the surrounding environment. The leaseholder is responsible for all operations including their effects on safety of people and property, and protection of the environment.

Crew Safety

OCS operations present a continuing risk of accident and injury. Drilling operations involve moving heavy equipment into place (e.g., pulling or hauling pipe) and the continual adjustment of controls and rotary equipment. These operations are personnel intensive. Production operations likewise involve the continual maintenance of process equipment, as well as activities associated with changing flowrates and reservoir depletion. All offshore operations also involve the lifting and moving of heavy loads and other manual tasks.

The workplace itself involves many sources of hazard. Offshore platforms have limited work areas, elevated walkways, and equipment installed in cramped spaces. Openings on the deck present crews with a risk of falls. Storms, wind, and lightning have been factors in workplace accidents. Well blowouts which can result in explosion and fire, as well as breaks and leaks in pipes or hoses, have been the source of most catastrophic accidents. The MMS defines as a "major accident" a fire or explosion that results in equipment or structural damage of $1 million or more, hydrocarbon spills of 200 barrels or more during a period of 30 days, or a fatality or serious personal injury (U.S. Department of the Interior, 1988a). There were a total of 206 fatalities reported over the 32-year period between June 1956 and December 1988. Seven fatalities occurred in early 1989, just before this report was completed.

Protection of Facilities and Property

The financial consequences of the loss of a platform and its onboard equipment as a result of a blowout, fire and explosion, storm damage, or a ship collision could amount to several hundred million dollars just in replacement costs. The construction and installation costs for Shell's Cognac platform (located in 1,025 ft of water) were $255 million when it was installed in 1978 (Cognac production facilities cost an added $35 million). While this facility still is one of the deepest production structures built, its cost is comparable to the costs for large facilities common to the Pacific Coast, which, although built in shallower water, have extensive processing facilities. Loss of production is also a major cost element. Fortunately the United States is not dependent on a few OCS production facilities as compared to Britain, whose North Sea production suffered a reduction of over 5 percent from the loss of the Piper Alpha production platform in 1988.

Environmental Protection

The environmental pollution record relating to oil production on the OCS has been good. Of some 5 billion barrels of oil produced between 1972 and 1986, only about 80,000 barrels were spilled (U.S. Department of the Interior, 1988b).[3] While over 24,000 wells were drilled in that period, no blowouts occurred that resulted in significant amounts of oil reaching shore, impacting sensitive environments, or causing loss of resources. Pollution prevention has been an increasingly important priority of the leaseholders and the MMS.

Accident and pollution statistics are examined in more detail in Chapter 2.

---

[3]Page 90, Table 64 (spills of 50 barrels or more) of U.S. Department of the Interior, 1988b. In 1987, 297 barrels were spilled that were not recorded in time to be included in this table. Note that accident event reports for 1972-1986 show that spills from major offshore pipelines accounted for over 47 percent (37,000 barrels) of the reported spillage (U.S. Department of the Interior, 1988a).

Conservation of Resources

The loss of oil or gas because of accidents to facilities has a financial impact and potentially an environmental effect, but equally important, these losses deplete domestic resources making the United States even more dependent on imported oil. Resource losses from accidents on the OCS have been low, particularly during the last 15 years as noted in the discussion about environmental protection above.

SUMMARY

Offshore platforms and drilling and production facilities on the OCS now tend to be older (increasing average age) and are becoming larger and more complicated. Platforms in the Pacific OCS are more complex (more onboard processing) but are relatively near shore, while those in the Gulf of Mexico are situated farther from shore, at an increasing average water depth as new areas of the Gulf are placed under lease. Increasing distance from shore makes each unit more isolated and reliant on its onboard resources. The distance from shore support of Gulf facilities also inhibits frequent inspections by regulatory agencies. The greater size and complexity of the Pacific platforms currently requires larger crews.

Through the years, the Gulf of Mexico facilities have provided over 95 percent of the United States offshore oil production. Over half of these facilities are more than 15 years old. Aging of production systems presents an obvious potential for operational malfunctions, both minor and catastrophic. The need for effective on-site safety programs is, therefore, becoming greater despite the good overall safety record of the last 15 years.

2

CURRENT INSPECTION PROGRAM: EVOLUTION AND PRACTICE

The current OCS platform inspection program was developed by the U.S. Geological Survey (USGS) of the Department of the Interior in the late 1960s and has not been changed significantly in its basic structure and approach since that time. The program became part of the Minerals Management Service (MMS) mission when it assumed the leasing administration functions of the USGS in 1981. The overall safety record on the outer continental shelf (OCS)–defined in terms of major accidents, deaths and oil spills–has been relatively good when compared with the safety record of the offshore oil and gas industries of some other nations. It is comparable to that of the most hazardous industrial activities ashore, such as mining and construction. The record concerning injuries is less clear since the data are imprecise and are not stated in standards common to records in other industrial operations. The problems of safety and environmental risks on the OCS have been studied thoroughly by the government, individual investigators, and contractors (NRC, 1981, 1984; MIT, 1986). Over the past 15 years measures have been developed to reduce those risks. Leaseholders recognize that they have an enormous economic stake in maintaining safe operations. The problem is to keep control devices in good repair and to maintain high levels of safety awareness over time.

This chapter describes the inspection program and practices currently employed by the MMS on the OCS. There follows a summary of the regulatory and technical underpinnings for the system of safety-related requirements in force today in drilling and production operations. Finally, the OCS safety record is examined from a statistical standpoint.

PRESENT INSPECTION PRACTICE

Overview of Inspection Program

All oil and gas operations conducted on the OCS are inspected periodically by the MMS. Inspection oversight is decentralized in the regions that administer the leasing program. The MMS is required under the OCS Lands Act (OCSLA) to provide for (1) scheduled "announced" on-site inspections, at least once a year, of each facility on the OCS which is subject to any environmental or safety regulation; and (2) periodic "unannounced" onsite spot inspections without advance notice. In its permit process MMS verifies the leaseholder's qualifications to operate safely and specifies the frequency of inspection and testing of safety systems by the leaseholder. It monitors drilling, production, and workover operations; testing of safety, pollution prevention, and metering equipment; and personnel qualifications. The MMS also monitors compliance with the provisions of the many environmental protection laws and their associated regulations and lease sale stipulations.

The present onsite inspection program of the MMS is built around a body of specific requirements promulgated by MMS, along with onsite monitoring for compliance carried out by an MMS inspection force.[1] Monitoring is facilitated by requirements that operators[2] perform tests, conduct drills, and maintain records that will afford MMS inspectors an overview of the status and history of installed safety devices and required functional drills. MMS regulations specify that safety devices to be tested and drills be performed weekly, monthly, semiannually, and annually on a scheduled basis. The records of the tests and drills are required to be kept on each platform or in the operator's nearest field office. Table 2-1 illustrates test activity by one sizable Gulf of Mexico lessee on its production platforms. Of the 78,813 tests performed during the year, 9.3 percent actually were witnessed by the MMS. The remaining 90.7 percent were performed by the operator without MMS supervision, but with the documentation being kept available for review by the MMS.

In order to expedite inspections and enforcement and to promote uniformity of inspectors' findings, all MMS standing requirements (derived from regulations) have been melded into a listing of "potential incidents of non-compliance" (PINCs), which forms a checklist for the MMS inspectors. This "National PINC List" generally addresses tests of safety devices. The underlying concept is that, in the design of facilities, the safety devices are critical for preventing accidents and pollution; that is, these devices are designed to "shut-in" and contain the system in the event of a failure, and therefore must function as designed.

A determination that there has been non-compliance with any item on the PINC list is termed an "incident of non-compliance" (or INC), and constitutes a violation of MMS regulations. The PINC list specifies the prescribed enforcement action. The more serious violations (INCs) elicit a shut-in order, which requires immediate corrective action and shutdown of the particular operation involved (or even the entire facility) until correction is effected. Other violations can result in civil penalties, but penalties have been imposed rarely since a court holding that operators must be afforded the opportunity to take corrective action before a penalty can be invoked. The great majority of INCs are deemed noncritical[3] and result only in a warning, subject to the operator taking prescribed corrective action within a specified time period. The first part of Appendix D (which is representative) lists INCs reported on two operators during an annual inspection; the second part is a partial biennial summary of the nationwide INC/PINC ratio for drilling and production during a recent year.

---

[1]The focus of this study is inspection related to ongoing operations. It should be noted, however, that initiation of both drilling and production operations involves a series of permit approvals directed to assuring that a facility is safe (i.e., that it has in place a plan for achieving safety performance goals, as well as effective safety procedures and equipment) before drilling or production operations can begin.

[2]OCS lessees are responsible for the safe operational performance of a facility. However, on a practical basis, an "operator" approved by MMS–who may be a sole lessee, one of a group of joint lessees, or a third party engaged by the lessee(s)–manages the day-to-day activity on a platform. Since this report deals with the safety of onsite operations all references are to the "operator," recognizing that ultimate accountability for safety performance lies with the lessee(s).

[3]The categorization "critical" varies in meaning across regions. The term generally relates to requirements directly affecting well control and is not limited to safety devices. The Alaska OCS Region makes no formal distinction between critical and noncritical.

TABLE 2-1 Testing Activity by an Operator/Lessee on Production Platforms

Platform No.	Potential Incidents of Non-Compliance (PINCs): Monthly checks[a] approx. no.	Quarterly checks[b] approx. no.	Semi-Annual checks[c] approx. no.	Annual checks[d] approx. no.	Failures or INCs[e] since last MMS annual inspection	Percentage INCs to PINCs
1	450	15	60	20	1	0.18
2	135	0	5	0	2	1.43
3	565	15	33	20	3	0.47
4	250	1	25	20	4	1.35
5	270	2	28	20	3	0.94
6	430	3	42	10	12	2.47
7	430	3	41	15	0	0.00
8	160	1	14	5	3	1.67
9	70	0	2	0	3	4.17
10	250	2	25	8	3	1.05
11	135	3	26	10	3	1.72
12	245	1	23	4	5	1.83
13	305	12	29	15	1	0.28
14	360	4	29	8	0	0.00
15	315	1	31	8	0	0.00
16	240	2	16	6	0	0.00
17	215	1	21	5	0	0.00
18	230	1	26	6	0	0.00
19	665	16	63	20	2	0.26
20	215	6	30	5	0	0.00
21	475	16	56	18	2	0.35
Company-wide results (# inspections by all parties)	6410	105	625	223	47	0.64
Per Year	x12	x4	x2	x1		
Total Inspections[f]	76,920	420	1250	223	47	

[a] Number of required Safety and Pollution Prevention Equipment (SPPE) items to be checked or verified monthly, (e.g.,pressure switches high/low (PSHLs), level switches high/low (LSHLs), shutdown valves (SDVs), flow safety valves (FSVs)

[b] SPPE items to be checked or verified quarterly (e.g., gas detectors)

[c] SPPE items to be checked or verified semi-annually (e.g., pressure safety valves)

[d] SPPE items to be checked or verified annually (e.g., temperature switch, high; burner switch, low)

[e] INCs are Incidents of Non-Compliance

[f] The aggregate total of inspections by the operator was 78,813 of which 7,363 were witnessed by MMS personnel.

Note: See Chapter 2 regarding development of SPPE list of items.

Inspection Policy

Basic policy guidance for lessees and inspectors is set out in the *Minerals Management Service Manual*, Chapter 650.1. (U.S. Department of the Interior, 1984) Highlights are (1) a reiteration of the statutory requirement making each facility subject to an annual (announced) onsite inspection–the manual specifies that all "applicable inspection characteristics delineated on the National PINC list" are to be inspected;[4] and (2) a reiteration of the statutory requirement making each facility subject to spot (unannounced) inspections–the manual specifies that a minimum of 10 percent of all production facilities are to be spot-inspected annually and that at least 25 percent of the applicable PINCs are to be witnessed (for drilling facilities the percentages are 50 percent of the facilities and 25 percent of the PINCs). Each facility also is subject to unannounced inspections of specific items and systems. Directly supporting the manual (Chapter 650.1) is the *Minerals Management Service Handbook* (U.S. Department of the Interior, 1986). This directive establishes the agency's inspection organization and authorizes promulgation of field office supplements covering actual methodology to be used in conducting inspections. These supplements are subject to concurrence at the headquarters level prior to promulgation by the regional director.

The handbook outlines procedures to be followed by inspectors in handling incidents of non-compliance, including notifications of operators, and for maintaining and submitting inspection records. It also describes internal review procedures designed to monitor a region's inspection process. The National PINC List was promulgated initially as a headquarters directive to bring national uniformity to the inspection process at a time when each region was enforcing locally promulgated "OCS Orders." (The MMS recently issued uniform regulations which supersede the regional OCS Orders; these regulations went into effect on May 31, 1988. The PINC lists have been revised to conform to the new regulations.)

Inspection Procedures

Inspections are conducted by district employees classified as Petroleum Engineering Technicians–their entry level is GS-9, journeyman level is GS-10, and supervisory level is GS-12–who are required to have had field experience in the oil and gas industry prior to being employed by the MMS. Appendix E presents a job description for the Petroleum Engineering Technician (journeyman level). For the most part, the technicians' (inspectors') prior qualifications reflect practical experience on a drilling facility and/or a production platform rather than formal training in inspection techniques. Generally speaking, specific training in inspection procedures is limited to on-the-job training gained while accompanying a trained technician.

To some degree inspectors are rotated with a view to maintaining an arms-length relationship with facility personnel.

The dispersion of the facilities requires that large amounts of working time be used in getting to and from facilities (notwithstanding the nearly total reliance on helicopters for transportation). This results in inspector time (for production operations) being utilized at a rate of about 70 percent in the Gulf and 59 percent in the Pacific Coast (MMS, unpublished information).

Once at a facility, inspectors customarily review the facility records of the tests and drills conducted by the operator to meet MMS requirements, as well as operational records. These reviews appear to be far from comprehensive, but nevertheless, are time-consuming. (Some review of MMS records is done ashore prior to the inspection to give the inspectors specific information

---

[4]The PINC lists differ for production facilities and for drilling facilities. The PINC list for production operations was derived in part from the American Petroleum Institute's Recommended Practice Standard 14C, which covers safety systems for offshore production platforms. The lists recently have been expanded to address workover operations.

regarding the inspection history of the facility.) Thereafter, the inspector tours the facility accompanied by a technician from the facility who is qualified to operate machinery and conduct tests required by the inspector. MMS personnel do not operate machinery or test equipment, but they do select equipment to be checked and tests to be conducted. The choice of items is viewed by inspectors as something to be determined "using their expertise," not only during unannounced inspections where selective witnessing clearly is authorized, but also sometimes during annual inspections, notwithstanding MMS's 100 percent inspection standard. The criteria used for this selection were not identified or discernible to the committee during onsite visits.

Inspector responsibilities in the Pacific region also include inspection tasks to ensure operator conformance with federal air emission standards. This activity presently affects 4 of 20 platforms, but it is expected to extend to all OCS facilities. A small portion of the time of one MMS engineer and two MMS technicians is dedicated to checking the volatile organic compound (VOC) emissions of gas turbines, monitoring water injection ratios, spot-checking exhaust temperatures, and reviewing maintenance records reflecting emissions performance. Other environmental responsibilities of inspectors in the Pacific region include investigation of flaring (gas burning), fuel consumption rates of equipment, and hydrogen sulfide concentrations and pressures.

At the conclusion of the inspection, inspectors advise the facility personnel of any deficiencies involving a warning; those involving a shut-in are addressed as they are observed. A record of the deficiencies is made and filed at the district office for follow-up purposes. The G-400 "safe and workmanlike manner" PINC is sometimes used as a way to flag dangerous conditions not covered elsewhere on the list.[5]

On each visit to a platform, the inspector reviews a summary of the previous inspection and checks to ensure that any INCs have been corrected. Most INCs requiring shutdown of an operation can be corrected while the inspector is still in the field. In such cases the inspector will authorize resumption of operations. Where the INC is corrected after the inspector leaves the platform, the operator will notify the district office, by phone, and receive verbal permission to resume operation. Depending upon the type of INC and the inspector's schedule, the inspector may or may not check to verify compliance prior to the next annual inspection.

Safety Communication to Operators

The MMS OCS regional or district offices issue "safety alerts," which are notices to operators and lessees about operational safety problems and recommendations for eliminating hazards or dangerous practices. Most safety alerts deal with circumstances that have contributed to or caused injuries. These notices are issued on an ad hoc basic, but are not necessarily related to an accident or fatality event, nor to analyses of INCs.

Summary

The complete PINC list is intended to lead to a comprehensive check of all systems and operations on OCS facilities so that a valid, instantaneous picture can be drawn of the condition of the listed systems and operations. MMS officials take the position that because process upsets resulting from human error, poor maintenance, and unexpected operational events can be sensed and controlled by safety systems, inspecting the system for compliance with PINC list requirements is a valid means of ensuring systems safety and integrity. In essence, then, the current inspection program and practices conform to the MMS's interpretation of public law as derived principally

---

[5]G-400: "Is each operation performed in a safe and workmanlike manner, and are the necessary precautions taken to prevent health, safety, or public hazards?"

from the OCSLA. The questions are: *Is this interpretation adequate from the standpoint of public expectation regarding safety on the OCS? Is there a different approach to inspection that would enhance safety in the long run?* These are questions that this report attempts to answer.

TECHNICAL FOUNDATION FOR INSPECTION REQUIREMENTS

One approach to evaluating the current inspection system is to examine its technical basis. The MMS inspects both onboard systems and operational procedures for compliance with safety requirements. The requirements for what should be inspected in drilling and production operations and the necessary safety devices were established by the USGS soon after federal regulation of OCS operations began. The requirements were based on technical findings and on formal safety analyses, and have been updated to a limited extent by the MMS as experience has indicated. This section briefly describes those technical bases.

Evolution of the Safety Practices for Drilling Operations

Requirements regarding procedures to be used to determine casing depths and mud weights, as well as procedures for kick detection and requirements for control devices, are imposed by MMS regulations carried over from the USGS. These requirements are based on the findings of informal reviews by various American Petroleum Institute (API) and industry committees, of previous blowouts and industry experience gained over several decades of operations offshore, and a longer history of land operations. The hazards of drilling and the equipment required to address them are well known. In essence, subsurface conditions dictate the type and amounts of casing and tubing, and the type of wellhead. This MMS safety approval of the design of newly drilled wells requires the installation of surface and subsurface safety valves.

Evolution of the Requirements for Safety Devices for Production Facilities

Detailed federal requirements for safety and pollution control devices were first promulgated October 30, 1970, in a revision of USGS OCS Order No. 8, dealing with the installation and operation of platforms, and OCS Order No. 9, pertaining to oil and gas pipelines. API's Offshore Safety and Anti-Pollution Equipment (OSAPE) Committee, and the Offshore Operators Committee (OOC)[6] provided technical advice to the USGS on the development of the requirements. Subsequently, the USGS commissioned contractors to perform safety analyses on several representative Gulf of Mexico facilities. Both Hazard Analysis (HA) and Failure Modes and Effects Analysis (FMEA) techniques were employed to develop a standard procedure for evaluating hazards at offshore production facilities and ensuring that appropriate safety devices were installed. After reviewing the contractor's work, the oil and gas industry pointed out that a modified FMEA approach was more compatible with the practices of the industry and would supply the necessary information in a manner that was both simple to gather and easy for the government to verify in its design review. The API then developed this concept into its Recommended Practice 14C, (*Analysis, Design, Installation and Testing of Basic Surface Safety Systems for Offshore Production Platforms*, referred to as RP 14C, first published in June 1974). This document soon was incorporated into the government regulations. It has been updated periodically.

---

[6]The Offshore Operators Committee is composed of representatives of the operating companies in the Gulf of Mexico.

The modified FMEA approach evaluates each piece of equipment as an independent unit, assuming worst-case conditions of input and output, including the failure of a device and its influence on the environment (e.g., gas, fluids) around other devices. Separators, flowlines, heaters, compressors, etc., function in the same manner regardless of the specific design of the facility. That is, they have fluid level, gas pressure, temperature controls, valves, and in some cases relief valves. These are subject to failure modes that impact the piece of equipment in a consistent manner. Thus, if a FMEA is performed on the item of equipment standing alone, it will be valid for that component in any process configuration.

Furthermore, once every process component has been analyzed separately for worst-case stand-alone conditions, there is no additional safety risk created by joining the components into a system (Arnold and Stewart, 1989). That is, if every process component is fully protected based on its FMEA analysis, a system made up of a number of these components also will be fully protected.

It even is possible to configure the system so that protection furnished by devices on one process component can protect others. That is, devices that may be required to provide adequate protection for a component standing alone may be redundant once all components are assembled in a system.

The RP 14C procedure for determining what safety devices are required to protect the system and process is summarized below:

- For each piece of equipment (process component), develop a FMEA by assuming that every process upset that could become a potential problem in fact occurs.
- Provide a sensor that detects the upset and shuts-in the process before an identified hazardous condition develops.
- Apply FMEA techniques to identify an independent back-up to the sensor that will provide a second level of defense before the identified hazard develops. The required degree of reliability of the back-up device depends on the severity of the potential hazard.
- Where experience in applying FMEA analysis to production equipment indicates that only one level of protection is required because of the degree of reliability of the primary device and the low cost of the specific consequences of failure, nevertheless, two levels of protection should be specified since it is often more costly in engineering time to document the need for only one level of protection than it is to install and maintain two levels.
- Assemble the components into the process system and apply FMEA techniques to determine if protection devices on some components provide redundant protection for other components.

Each installation design submitted for MMS approval must contain a process schematic and a Safety Analysis Function Evaluation (SAFE) Chart, which lists each process component and the devices required by the modified FMEA. If a device is installed, the chart indicates how the device acts to isolate the component (e.g., which valves it actuates). If a device is not included, an alternative device that will provide the protection is recorded. The combination of process flow diagram and SAFE chart, along with the feedback of data from inspections, ensures that the MMS design review for safety devices is comprehensive.

SAFETY PERFORMANCE RECORD

The safety of OCS operations can be measured in terms of accidents and resulting deaths and injuries, as well as pollutant (primarily oil) spillage into the environment. At the extreme are losses of entire platforms due to blowouts, explosions, and fires.

Any discussion of the safety performance record on the OCS must make note of the fact that OCS data collection currently has many shortcomings and that data collected by the MMS and the

U.S. Coast Guard are not consistent. For example, data about deaths and injuries in a given year may be quite different. Several variables contribute to the discrepancy:

- Reporting requirements and procedures for the two agencies are different.
- The number and types of offshore platforms required to be reported are different.[7]
- The reporting year itself is different (i.e., calendar year vs. fiscal year).

The committee had available to it parallel statistics compiled by both MMS and USCG. Because its primary focus is on MMS inspection practices, the committee has relied primarily on MMS data, referring to Coast Guard figures where pertinent.

Adequacy of MMS OCS Safety Information System

In two studies, the Marine Board conducted a comprehensive review of the technology and regulations relating to OCS oil and gas development (National Research Council, 1981, 1984). The committee that conducted the 1981 investigation found that "without a strong safety information component in the OCS regulatory program, it is not readily possible for the government to identify safety problems and courses of action." The 1984 study recommended that the MMS establish an OCS Safety Information System for monitoring the safety performance of OCS owners and employers. This system should (quoting)

- acquire comprehensive OCS event and exposure data;
- relate events to specific employers, locations, operations, and equipment;
- calculate frequency and severity rates and analyze trends; and
- permit monitoring of the relative safety performance of owners and employers, locations, and activities.

Part of such a system has been developed and put into effect, but the reliable safety information system essential to the management of MMS inspection operations and effective safety management still is not available. Some of the data are poorly organized or incomplete; in particular worker exposure information is lacking. Accordingly, there is an inadequate basis for

- comparison of the safety performance of OCS operators with operators in other industries,
- detection of time-related trends in safety performance,
- identification of particular safety problem areas in OCS operations that may require special attention, and
- evaluation of the effectiveness of new or changed inspection requirements.

Injuries and Fatalities

A primary aim of the MMS inspection program is to ensure that OCS operations do not expose operating personnel to unwarranted physical hazards. In order to assess the relative risks in various industries, it is useful to compare injury and fatality incidence rates. The Bureau of Labor Statistics (BLS) of the Department of Labor, and other federal agencies such as the Bureau of Mines (BOM), publish statistics concerning incidence rates of fatal and nonfatal occupational

---

[7]The Coast Guard separates data relating to MODUs and fixed platforms; these data cover MODUs in state waters and U.S.-owned units overseas as well as those on the OCS.

injuries and illnesses, lost workdays, etc. Such statistics are widely available and can be used for industry-to-industry comparisons and for trend analysis.

The only readily available public record of fatal and nonfatal injuries resulting from OCS operations is a compilation published by the MMS entitled, *Accidents Associated with Oil & Gas Operations* (U.S. Department of the Interior, 1988a). This report supplements the previously published *Accidents Connected with Federal Oil and Gas Operations on the Gulf of Mexico Outer Continental Shelf*, Volume I (OCS Report MMS 86-0038) and Volume II (OCS Report MMS 87-0049) and updates the data contained in those publications. The scope of the new report is broader; it covers the entire OCS during the reporting period. Table 2-2 sets out injury and fatality records for the period from 1977 to 1986, inclusive, extracted from the MMS reports and partial data for 1987 acquired by the committee. This table includes injuries and fatalities in the Gulf of Mexico and Pacific regions. The Atlantic and Alaska regions have "not had any accidents that meet the criteria set forth in the Introduction" to the report U.S. Department of the Interior, 1988a.

The accidents involving human deaths and injuries listed in the MMS report are those associated with blowouts, explosions and fires, significant pollution incidents, and major accidents. No definitions of "injury" and "fatality" are given in the MMS report, apart from a statement that defines major accidents as those involving–among other things–"all fatalities or serious personal injuries that cause impairment of any bodily unit or function" (U.S. Department of the Interior, 1988a).

The U.S. Coast Guard provided the statistics for injuries from all causes for the three-year period from 1984 to 1986, as listed in Table 2-2. The Coast Guard also records data on injuries related to equipment failures. The number of injuries from all causes in the Coast Guard data is

TABLE 2-2 Injuries and Fatalities on the OCS, 1977-1987

	MMS data		USCG data	
Year	Injuries[a]	Fatalities[b]	All injuries[c]	Equipment-related injuries[c]
1977	27	4		
1978	12	9		
1979	15	17		
1980	83	28		
1981	22	12		
1982	45	18		
1983	29	8		
1984	30	19	873	22
1985	32	14	851	29
1986	10	7	454	13
1987	NA	1	NA	NA

[a]Data from Tables 2-A, 2-B, 2-E, and 4-A, 4-B, and 4-E (overlaps have been eliminated) and MMS Regional Office data information (1987 partial data).
[b]Data from Tables 2-E and 4-E.
[c]USCG correspondence to committee.

NA = Not available

more than 26 times as large as those reported by the MMS (the ratios are 29.1, 26.6, and 45.4 for the years 1984, 1985, and 1986 respectively). It is unlikely that the difference is due to the differences in the definition of injury.[8] A more likely explanation is that the majority of accidents resulting in personal injury in the Coast Guard data are defined differently than those categorized and reported by the MMS as "major accidents."

Exposure and Rates of Incidence of Injuries and Fatalities

No report is available from public agencies concerning the aggregate exposure of personnel on OCS facilities to safety hazards. As comparison with other industries is possible only on the basis of rates of incidence, a request was made to the Coast Guard to provide the committee with estimates of the daily population at risk on offshore facilities.

The Coast Guard supplied an average daily population estimate for the seven-year period between 1980 and 1986, inclusive. This estimate was based on the known number of structures and vessels of various kinds and on the estimated average crew size for different operations. A summary of the data supplied is given in Table 2-3.

TABLE 2-3 Estimated Daily Population of Workers on the OCS, 1980-1986

(i) Personnel on mobile offshore drilling units (MODUs)

Type of structure/vessel	Average number per day
Submersibles	732
Semisubmersibles	1,200
Drillships	399
Jackups	4,436
Subtotal	6,767

(ii) Personnel on manned platforms

Type of activity	
Production operations	10,385
Drilling operations	3,784
Subtotal	14,169

(iii) OCS personnel

Total	20,936

SOURCE: U.S. Coast Guard, unpublished data, 1988

[8]As contrasted to the MMS, which reports injuries only if they are related to accident "events", the Coast Guard uses a more specific definition of injury (see following subsection) that excludes only minor cuts, bruises, and burns, as well as time off the job.

In calculating the total hours of exposure it must be appreciated that offshore facilities operate 365 days per year. Personnel generally work a 12-hour shift, then spend 12 hours off duty. However, since the workers are domiciled on the facilities for the full day, they are exposed to certain hazards even when they are off duty, which is not the case in land operations. In these circumstances, the committee has deemed it necessary to consider both working (12 hr/day) and total (24 hr/day) exposure.

Accordingly, the total hours of exposure per year were calculated as follows:

- Working hours of exposure (EH) for a calendar year (while on duty):
  EH_{wkg} = 12 x 365 x 20,935 = 91,669,680 hours/year
- Total hours of exposure for a calendar year (while domiciled, whether on duty or not):
  EH_{tot} = 24 x 365 x 20,935 = 183,399,360 hours/year

The number of injuries reported by the Coast Guard and included in Table 2-2 is based on the definition of injury used in the CASMAIN Data Base (U.S. Coast Guard Casualty Reporting Requirements, 33 C.F.R. 146.30). There, a reportable injury is defined as injury to five or more persons in a single accident or an injury causing any person to be incapacitated for more than 72 hours.

Unfortunately, this definition differs from that used by the Occupational Safety and Health Administration (OSHA), which gathers accident data for many industries. OSHA categorizes reportable nonfatal occupational injuries into two groups–see, for example, BLS Bulletin 2259 (Bureau of Labor Statistics, 1986):

1. Cases involving lost workdays which are either lost workday cases involving one or more days away from work, or lost workday cases involving one or more days of restricted work activity only.
2. Cases without lost workdays that result in transfer to another job, require medical treatment (other than first aid), or involve loss of consciousness or restrictions of work or motion.

Neither of these OSHA definitions is sufficiently close to that used by the Coast Guard to justify a comparison of injury incidence rates. This finding is further illustrated by the data in Table 2-4. The ratios of number of injuries to number of fatalities are much greater for industries reported on by the Bureau of Labor Statistics, (BLS, 1986) than the ratio obtained for OCS operations from the Coast Guard reports. The lowest (worst) ratio seen on the BLS data is for mining, a ratio (139) that is still 2.6 times greater than the ratio for OCS (54) (Table 2-3). The likelihood of having such a large discrepancy, if the two sets of data were comparably based, is small.

The OCS fatality incidence rate is calculated from the formula (N/EH) x 200,000,000 where N is the number of fatalities, EH is the total hours worked by all employees during a calendar year, and the constant is the standard used by the Bureau of Labor Statistics (2,000 labor hours per year x 100,000 employees) for reporting incidence rates. A minimum (working) and a maximum (total) value of EH in OCS operations was calculated earlier. According to Table 2-2, the average number of fatalities during the seven-year period from 1980 to 1986 (the same seven years used by the Coast Guard to estimate daily population at risk–see Table 2-3) is (106/7) = 15.1 fatalities/year. On this basis, the maximum and minimum fatality incidence rate for OCS operations are 16.5 and 33.0, respectively. To bring these figures into perspective, the fatality incidence rates in various industries in the private sector are given in Table 2-5, for 1983 and 1984 (Bureau of Labor Statistics, 1986). Clearly, OCS operations result in fatality incidence rates which are comparable with those seen in the more hazardous industries in the United States (e.g., mining and construction). This is not an unexpected result, given the work environment prevailing on the OCS.

TABLE 2-4 Ratios of Injuries to Fatalities, 1984[a]

Industry	Injuries (lost workday cases)	Fatalities	Injuries/ fatalities (ratio)
Agriculture, forestry, and fishing	46,300	110	421
Mining	51,400	370	139
Construction	256,500	660	389
Manufacturing	841,800	800	1,052
Transportation and public utilities	249,300	770	324
Wholesale and retail trade	574,300	440	1,305
Finance, insurance, and real estate	44,300	80	544
Services	385,800	510	756
Private sector total (non-OCS)	2,449,700	3,740	655
OCS Oil and Gas[b]	2,178	40	54

SOURCES
[a]BLS, 1986 for all data except those for OCS.
[b]Table 2-2, covering years 1984, 1985, and 1986.

TABLE 2-5 Fatality Incidence Rates

Industry	1983	1984
Agriculture, forestry, and fishing	12.7	16.3
Mining	27.6	41.4
Construction	26.3	22.8
Manufacturing	4.3	4.4
Transportation and public utilities	13.3	16.9
Wholesale and retail trade	3.3	3.1
Finance, insurance, and real estate	1.7	1.9
Services	2.2	3.9

SOURCE: BLS, 1986.

It must be noted however, that since the early 1970s no major disasters have occurred on the U.S. OCS that approach the scale of recent disasters on platforms in the North Sea. For example, the July 1988 Piper Alpha explosion and fire cost the lives of 166 of the 229 people on board.

This country's OCS operations do not have the concentration of people found in the North Sea operations. However, the possibility of an accident occurring on the OCS that could result in 50 to 100 fatalities cannot be dismissed. A single occurrence of this magnitude would have caused the fatality incidence rate over the seven-year period 1980 to 1987 to have doubled.

Causes of Fatalities and Injuries

Table 2-6 presents the committee's analysis of the causes[9] of deaths and injuries in the 54 events that produced fatalities in the Gulf of Mexico over the period 1982 to 1986, as recorded in the MMS's events file (U.S. Department of the Interior, 1988a). The handling of heavy loads and falls were the two most common sources of these accidents.

TABLE 2-6 Causes of Accidents Involving Fatalities in the Gulf of Mexico, 1982-1986

Description of cause	Number of events	Number of fatalities	Number of injuries[a]
1. Handling heavy loads (including crane accidents)	16	16	5
2. Fall from height	11	11	1
3. Opening a pressurized system for maintenance	7	8	3
4. Drowning	5	5	0
5. Illness and heart attack	3	3	4
6. Premature firing of perforating gun	2	2	4
7. Electrical shock	2	2	0
8. Vented gas during well work	2	6	4
9. Helicopter	1	2	2
10. Welding operations	1	3	9
11. Vapors from drain lines	1	1	1
12. Jackup rig failure	1	1	0
13. Unknown	2	2	2
Total	54	63	31

[a]Nonfatal injuries associated with fatality events.

[9]Causes as reported in the MMS events file.

Pollution (Spill) Incidents

A further hazard deriving from offshore gas and oil operations is the spilling of pollutants which may endanger the marine and coastal environments. MMS files and reports, using data reported by operators, are the original source of information concerning the frequency and volume of such spills from offshore facilities. The most up-to-date report is U.S. Department of the Interior, 1988a.

Data concerning oil spills of 50 barrels or more that occurred during the last decade are summarized in Table 2-7 for the Gulf of Mexico. The Alaska and Atlantic regions did not experience incidents that were reportable under the criteria used in the MMS report. The Pacific Region oil spillage was estimated to be about 1,500 barrels per year or less; this volume was attributed almost entirely to seepage, including natural seepage.

The pollution incidents in the Gulf of Mexico represent spill rates ranging from a low of 0.3 barrels per million barrels produced (or 0.00003 percent) in 1986, to a high of 20.6 barrels per million (or 0.0021 percent) in 1981.

While major spills can and do occur and produce widespread ecological damage and involve large cleanup costs, as in the case of the Ixtoc-I spill in Mexican waters (1979 to 1980), oil pollution from offshore operations contributes less than any other source to the release of hydrocarbons into the marine environment. A National Research Council assessment of worldwide inputs from natural sources (e.g., seeps), transportation (e.g., tanker operations and accidents, bilge oil, etc.), the atmosphere (particulates), municipal and industrial wastes and runoff, and offshore oil and gas production ranked the latter category lowest, at 0.05 million metric tons per annum (mta) of the 3.2 mta total (National Research Council, 1985). Refinery input (0.1 mta) and municipal wastes (0.7 mta) are sources that exceeded the average annual input from offshore operations. The same report noted that U.S. offshore spillage probably contributes less than 5 percent of the world offshore spillage. The data indicate that since 1977 there has been no obvious long-term trend in spillage rates from offshore oil and gas operations. However, over the last five years of record there has been a noticeable reduction in average spill volumes and in the amount spilled in comparison to production.

TABLE 2-7 Oil Spills of at Least 50 Barrels (bbl), Gulf of Mexico OCS

Year	Events	Total vol. (bbl)	Average spill vol. (bbl)	Production vol. (million bbl)	Oil spilled (bbl)/ oil produced (bbl/mill)
1977	4	670	168	291.7	2.30
1978	3	1,139	380	280.2	4.06
1979	5	2,095	419	274.6	7.63
1980	9	2,581	287	267.2	9.66
1981	6	5,562	927	270.2	20.58
1982	3	842	281	292.8	2.88
1983	9	1,939	215	317.8	6.10
1984	2	150	75	340.0	0.44
1985	8	1,354	169	359.5	3.77
1986	1	119	119	360.0	0.33
Totals (10 yrs)	50	16,451	329	3,054	5.39

SOURCE: U.S. Department of the Interior, 1988a.

Of course, an historical analysis cannot exclude the possibility of catastrophic spills such as occurred in four events between 1967 and 1970 with spills of 10,000 barrels or more. These events spurred industry and government action, which appears to have enhanced the industry's capability to assure well control and process control.

Fires, Explosions, and Blowouts

One important measure of safety practices is the number of fires, explosions, and blowouts that occur aboard OCS facilities. Table 2-8 shows the number of such accidents occurring on the OCS since 1970 in the Gulf of Mexico (over 95 percent of OCS drilling activity). The number of incidents per year shows no clear trend even when compared to indicators of drilling activity, such as the number of new wells drilled per year (loss of well control during exploratory and development drilling is the leading cause of fires, explosions, and blowouts.)

Violations Cited During Inspections

Table 2-9 indicates that the average number of inspections per platform (production) has fallen slightly (e.g., a decrease of 18 percent–from 1.7 to 1.4–over 3 years) compared to those

TABLE 2-8 Major Accidents in the Gulf of Mexico OCS Region[a]

	Fires and explosions	Blowouts
1970	7	2
1971	1	2
1972	0	1
1973	3	2
1974	0	1
1975	4	2
1976	0	1
1977	3	0
1978	7	2
1979	3	1
1980	7	4
1981	3	1
1982	8	2
1983	4	3
1984	6	1
1985	3	1
1986	1	1

[a]All fires or explosions that result in equipment of structural damage of $1 million or more. Both production and drilling related events are included in the number. Note: A single accident event may be included under more than one accident category.

SOURCE: MMS, unpublished data.

TABLE 2-9 Drilling and Production Inspections and INCs

	Drilling				Production			
Year	No. active facilities[a]	Inspections	Average MMS inspections per facility	INCs	No. active platforms	Inspections	Average MMS inspections per platform	INCs
1985	250	5204	21	879	3373	5830	1.7	4017
1986	115	3143	27	570	3435	5282	1.5	3117
1987	145	2472	17	341	3540	4791	1.4	3224

[a]Average of low and high for year

SOURCE: MMS, unpublished statistics.

involving drilling activities, which showed a decrease of 53 percent over the same period when offshore exploration was curtailed greatly due to falling oil prices. However, the data do not reflect the scope of the individual inspections, which range from a full review of all PINCs to merely a brief visit to check the operator's resolution of an INC that had resulted in a shut-in.

Differences in the geographic distribution of platforms, the complexity of the facility (types of processing systems and number of wells), and the regional emphasis on environmental pollution are factors that may influence the number of assigned inspectors (see Table 2-10). These factors appear to be reflected in the inspection results shown in Table 2-11: for example, the INCs-per-inspection ratio in the Pacific compared to the Gulf of Mexico. However, it is not possible to infer from the available statistics the influence of each factor noted above on inspection frequency.

A point of major interest to the committee was the degree to which the absence of INCs (i.e., compliance with the PINC List) ensures safe operations. Table 2-12 gives the results of the committee's analysis of the 54 fatality-producing accidents listed in Table 2-6. In Table 2-12 the cause of each accident is attributed to one or more possible PINC violations. Only in 5 of the 54 accidents, involving 11 of the 63 fatalities, is a violation of a specific hardware-oriented PINC implicated. The rest can be related in a general way only to the PINC G-400. This PINC asks the question, "Is the operation performed in a safe and workmanlike manner?" It is described informally by MMS as a "catch-all" PINC, since it is not specific and is sometimes used by inspectors to cite poor operational procedures and general housekeeping. However, the committee observed that very few G-400 PINCs are issued by inspectors.

TABLE 2-10 Number of OCS Inspectors by Year, 1985-1987

Year	Gulf of Mexico	Pacific
1985	55	10
1986	54	11
1987	55	13

SOURCE: MMS

TABLE 2-11 INCs and Inspections in the Gulf of Mexico and Pacific

	Gulf of Mexico			Pacific		
Year	INCs	Inspections	INCs/inspections (ratio)	INCs	Inspections	INCs/inspections (ratio)
1985	4616	10120	0.45	406	2188	0.19
1986	3512	8401	0.42	242	2023	0.12
1987	3405	7404	0.45	238	1925	0.12

Note: The total number of inspections cited in Table 2-11 is greater than in Table 2-9 because workover, abandonment, metering (production verification), and pipeline inspections are included in Table 2-11 totals as well as production and drilling inspections.

SOURCE: MMS

TABLE 2-12 Analysis of Reported Accidents Involving Fatalities in the Gulf of Mexico, 1982-1986

Reported cause	Number of accidents	Number of fatalities	Number of injuries	Potentially violated PINCS
1. Handling heavy loads (including crane accidents)	16	16	5	G-400
2. Fall from height	11	11	1	G-400
3. Opening a pressurized system for maintenance	7	8	3	G-400
4. Drowning	5	5	0	G-400
5. Illness and heart attack	3	3	0	NONE
6. Premature firing of perforating gun	2	3	4	G-400
7. Electrical shock	2	2	0	G-300, G-400
8. Vented gas during well work	2	6	4	G-400, G-408 (?)
9. Helicopter	1	2	2	NONE
10. Welding operations	1	3	9	G-200, G-201, G-206, G-209, G-210, G-212, & G-400
11. Vapors from drain lines	1	1	1	G-400
12. Jackup rig failure	1	1	0	G-400
13. Unknown	2	2	2	UNKNOWN
Totals	54	63	31	

Thus, if possible violations of the "catch-all" PINC G-400 are disregarded, most of the accidents would have occurred even though the operator might have been in 100 percent compliance with all other PINCs at the time of the accident. The hardware-oriented PINCs seemingly are not tied closely to current accident experience–this observation does not imply that fatalities have not been averted as a result of these PINCs being enforced–rather it is meant to suggest that the safety-enhancement limits of a hardware oriented inspection program are being reached. PINC G-400 can be useful in addressing operational and procedural hazards, and indeed its increased use is being emphasized by MMS management. But, because it is so general in nature, it alone is not a sufficient tool to use in addressing operational hazards in a consistent and directed manner (see Chapter 7).

3

MAJOR CONSIDERATIONS IN FORMULATING AN INSPECTION PROGRAM

RELATIONSHIP OF INSPECTION TO SAFETY

Offshore drilling and production rights are granted to presumably experienced and competent operators following approval of the operator by the MMS. The public has a right to expect that such facilities will be operated with due regard for the safety of life, property, and the environment. The MMS inspection program was established to further this end. Nevertheless, the primary responsibility for safety lies with the operator. This is both an explicit and implicit condition of his lease. Therefore, the thrust of the MMS inspection program should be oversight to ensure that the operator has established and is following practices that will result in safe operations. This is different from the MMS undertaking to inspect every piece of equipment for proper functioning and give it a "seal of approval."

Accepting oversight as the thrust of the MMS inspection program, it is pertinent to explore the relationship of inspection to safety. This chapter of the report examines the issue to highlight different approaches to thinking about safety that the committee believes will be useful in evaluating alternative inspection programs.

Compliance versus Performance

A key question is, "What is the actual relationship between inspection and the safety of offshore platforms?" It is a truism that inspection contributes positively to safety, but it is widely accepted by safety professionals that too much inspection, the wrong kind of inspection, or the wrong attitude about inspection can detract from safety. A related point is the maxim, often cited in the manufacturing sector, that quality and reliability cannot be "inspected in." At best, inspection can only verify the quality and reliability that has been "built in." Thus, if designers and manufacturing engineers were to rely too much on inspection to catch built in defects, then the quality of the product inevitably declines, along with the spirit of the manufacturing organization.

Similarly, the committee believes that inspections will not ensure the safety of an offshore platform if proper equipment, operational procedures, and training have not been built-in. Inspection can only provide confidence that the safety that has been built-in is being realized. Inspection can provide confidence that effective procedures are being followed to ensure the proper functioning of equipment (particularly safety devices) and that the operating practices and safety attitudes[1] of the operator reinforce the designed safety measures. With too much reliance on inspection however, there is a danger that complacency can eventually set in, and that the goal of operators can subtly shift from "operating a safe facility" to "passing the safety inspection." When this happens, the resulting outlook could be termed a "compliance mentality," one that is preoccupied with satisfying regulatory requirements, as opposed to a "performance mentality," which focuses on safe and effective operation. The performance mentality regards compliance with regulations and passing inspections as an incidental and natural by-product of safe operations. The compliance mentality equates safety to satisfying the regulations, i.e., passing inspections.

Evidence of the compliance mentality has been seen often in many industries. For example, it is seen when an automobile manufacturer defends an unsafe location of a gas tank by saying that this location satisfies all applicable government requirements. It is seen in the launch of an Atlas-Centaur rocket during an electrical storm, its destruction by lightning, and the subsequent explanation that the launch criterion (viz., that visible lightning strikes must be at least five miles away) had been met.

The compliance mentality essentially is an abandonment of responsibility and judgment on the part of the operator and the passing of that responsibility to the regulating agency and its inspectors. In the case of OCS platforms a failure to recognize this responsibility can lead to an attitude on the part of the OCS operator that "This may not be the best equipment/procedure/decision, but it does satisfy the inspection requirements; therefore, if something goes wrong I am blameless." In much the same way, an inspector may fail to look for blatant safety problems because he has come to believe that, "My job is only to verify that all items on the checklist are OK. As long as I've done that job well, I can't be criticized if something goes wrong."

These attitudes have been stated in an extreme fashion, but they can arise at different times and to different degrees among operators and inspectors in any industry. The compliance mentality can corrupt both operators and the inspection organization. It can confuse the tool of the inspection–namely, following the PINC list–with its objective, which is improved safety. It insinuates a notion that "compliance equals safety." But no checklist can be completely comprehensive, nor can any inspector be completely knowledgeable about a facility. Indeed, those who live and work aboard the facility generally have a better understanding of its history, quirks, and problems, and thus a more accurate sense of where the hazards lie, than a visiting inspector with a checklist.

Clearly it is necessary to have regulations, inspections to enforce the regulations, and checklists to facilitate the inspections. The inspections and checklists must be updated regularly and modified to reflect emerging problems. But when they become the end rather than the means, the ability of operators and inspectors to make intelligent, responsible judgments is weakened, and safety is threatened.

How Much Inspection?

If more inspection does not necessarily equal greater safety, then how much inspection is the "right" amount–that is, how much will optimize the operator's safety performance? Figure 3-1 illustrates the question graphically. The illustration, while generic and simplistic, facilitates discussion.

---

[1]The committee defines a safety attitude as being "a state of mind, forced or otherwise, which is receptive to the generation of criteria or elements known to be associated with good safety performance." Elements commonly associated with a safe offshore working environment are written workplace procedures, regular inspections, safety meetings, training programs, protective clothing, environmental protection program, accident/incident investigation procedures, emergency response procedures, fire fighting capability, and training and life-saving equipment.

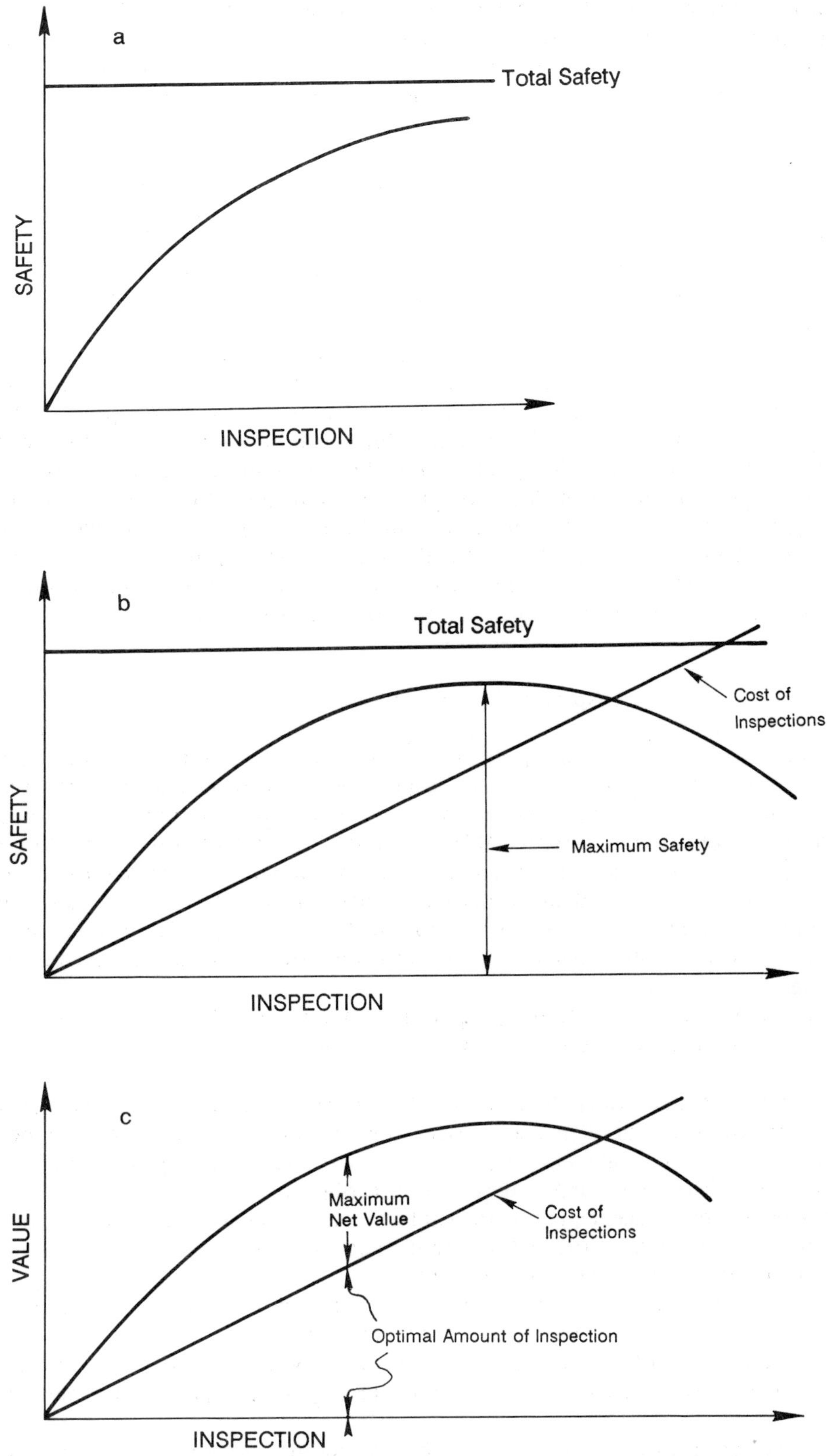

FIGURE 3-1 Relation of safety to inspection.

Figure 3-1a plots the amount of inspection against the degree of safety attained. It suggests that, as inspection increases in both thoroughness and frequency, safety also increases and approaches a maximum value, which we might call "total safety," in which preventable accidents never occur. However, there is a saturation effect, or a "law of diminishing returns," at play between inspection and safety performance so that additional increments of safety are increasingly difficult and expensive to obtain. This is the "first-order" part of the relationship.

Figure 3-1b shows a second-order effect in the relationship. As the amount of inspection is increased further, the degree of safety performance obtained from the operator actually begins to decline. This illustrates the compliance mentality effect described earlier. The point of "maximum safety" is therefore reached at a certain level of inspection, which is less than an "all out" effort. Figure 3-1b also includes a line representing the cost of inspection–both the direct cost to the government and the direct and indirect cost to the operator.

In Figure 3-1c, units of "Value to Society" are substituted on the safety axis to reflect the total cost savings of accidents avoided, including the societal value of lives saved and productivity maintained. The difference between the value curve and the cost curve now represent the net societal benefit gained from inspection. The greatest difference between these curves then represents the maximum net benefit. The point where this maximum is reached represents the optimal amount of inspection, even if it does not provide "total safety," or even "maximum safety."

Conceptually, therefore, there exists an optimal amount of inspection. Inspection activity is not a case of "more is better." To find that optimal point in practice, of course, is not a mathematical exercise. In practice the optimal amount of inspection must be established by trial and error, by experience and judgment, and by analysis of results over time.

Inspecting for Operational Safety

Accepting the premise that at some point more inspection does not necessarily yield greater safety, and that there is some optimal amount of inspection that provides society with the greatest return on its investment, there still remains the question of the optimal content of inspections. The PINC list is hardware-oriented, and current MMS inspection practice requires inspectors to concentrate on witnessing many tests of equipment. But, historically, as was described in the section on "Safety Performance Record," accident events have tended to be related to human error and to poor operating and maintenance practices, rather than to failures of specific devices. A more effective inspection program would therefore focus more intently on human factors in place of some of the hardware tests.

In regard to the area of human factors; experience in various industries indicates that operational safety can be increased by improving the following:

- Management attitude–when the attitude of supervisors and operating management encourages safety consciousness, accidents decrease. This attitude is reflected in awareness programs and individual discussions of the safe way to perform an operation, shutting in production where necessary, and setting realistic time targets to accomplish objectives.
- Housekeeping–many accidents are traceable to poor housekeeping. Where oil spills are cleaned up, equipment is properly protected and lubricated, parts and tools are neatly stored, and trash is properly handled, accidents are less frequent.
- Training–accidents decrease when workers are formally trained in the safe manner to perform the operations required of them, and in the correct use of tools.
- Communication–in a safe operation, near misses as well as accidents are analyzed and discussed. Workers are encouraged to raise questions about potential hazards in safety meetings.

Even brief observation of operations at a facility in any industry can give a clear impression of the degree to which attention is paid to these aspects of operational safety. The conscientiousness and general competency of workers, cleanliness and orderliness of the facility, and communicativeness of operator personnel are all readily apparent to the alert observer. To a large extent the impressions will be subjective and describable only in a relative sense (e.g., excellent, good, average, below average, nonexistent), but through the eyes of a properly trained inspector they can provide important information.

If MMS inspections are to enhance safety on the OCS they must encourage proper attitudes toward safety. The present MMS inspection program does not include a structured qualitative assessment of the four attributes listed above. Evaluations by several inspectors (especially during spot inspections) over a period of time could be used to establish a profile of all the operators on the OCS as well as for each individual facility.

Given the specific hardware orientation of the PINCs and the lack of a structured human factors assessment system, MMS inspectors might find it difficult to issue an INC when a facility is deficient in one of the attributes listed above (although such deficiencies could be addressed under PINC G-400). However, even a post-inspection discussion of deficiencies with the supervisor could be productive. Once a profile was established for a facility, the MMS could review the profile with the operator's manager responsible for auditing the operator's safety program and the performance of the supervisors. MMS should also mount programs to provide positive motivation for operators to "think safety" that are broader and more publicized than the SAFE recognition program currently in place.

The question arises as to how to handle the operator or supervisor who does not correct such deficiencies. The imposition of fines and penalties based on subjective judgments would be extremely difficult. As was often stated to the committee, the MMS cannot "legislate that individuals change their attitude." However, the operator is responsible for safety, and his response to human factors safety deficiencies could be encouraged through direct action such as shut-in of equipment located in areas where poor housekeeping is chronically evident, or equipment that workers have not been adequately trained to operate. More wide-ranging and frequent invocation of PINC G-400 is also an option.

ADVANCES IN SAFETY RISK ASSESSMENT AND MANAGEMENT

Over the past two decades a set of methodologies and techniques has emerged designed to assist system designers, developers, and managers in assessing and managing the safety risks of any complex system. This new activity, generally known as "risk assessment and management," is employed explicitly within a number of private industries and public enterprises engaged in the development and operation of complex systems. The nuclear power, chemical processing, and commercial aerospace industries are examples in the private sector; the U.S. Air Force is a public-sector example.

Risk assessment consists of identifying potential failure scenarios associated with the system, and assessing quantitatively the probability of occurrence of such scenarios. Risk assessment is an engineering function analogous to stress analysis. The results of risk assessment are inputs to the risk management function, whose objective is to make optimal design and operating tradeoffs between costly risks and benefits, and to institute a system of quality assurance or quality control to ensure that these are decisions implemented.

In the industries noted above, risk assessment is a formal process carried out by safety engineers, systems engineers, operations analysts, and other specialists using a variety of mathematical, statistical, and analytical tools. (Some of the most widely used are probabilistic risk assessment, failure modes and effects analysis, hazard analysis, and trend analysis.) The safety specialists report to line managers who are attuned to the significance and value of the analytic results. What was once a very uncertain process of guesswork and supposition has become an accepted, precise science that increases the confidence with which complex, technology-intensive systems of many kinds are designed, built, and operated.

The committee believes that some elements of this discipline are applicable to oil and gas drilling and production operations on the OCS. At the simplest level, the tools do not require extensive professional training to use. They can be employed–after some familiarization–by nonspecialists to identify in an objective way the items of equipment, operational procedures, facilities, and operators that should be accorded inspection priority. They can also be used to verify regulatory and inspection requirements and to discover failure modes that have not previously been recognized. The value and utility of these tools is enhanced by the collection of suitable types and amounts of data regarding the safety and operating history of OCS facilities. (The current MMS Offshore Information System [OIS] data system collects only some of the requisite data.) The limited data base on the performance of OCS systems available to undertake risk analyses was noted in a Massachusetts Institute of Technology (MIT) review of risk analysis in offshore safety and environmental management, which covered U.S., British, and Norwegian government policy (MIT, 1986). This report pointed out that most applications of risk analyses have focused on structural design, onboard oil processing, and fire safety concerns, but not on platform operations.

Chapter 7 elaborates on how the activity of the existing MMS inspection force can be redirected to utilize some of these tools and techniques to improve the inspection program.

CONFORMANCE WITH FEDERAL LAW

Apart from the adequacy of an inspection program in promoting safety, and the availability of new approaches to risk assessment and management, another important consideration relates to how well the inspection program and the regulatory process that it serves conform to federal law.

The broad mandate of the OCSLA implies that the MMS has the responsibility to protect against any events or conditions involving a serious threat to life, property, and the environment. In this context, compliance with the literal requirement of the OCSLA relating to a mandatory annual inspection and unannounced spot inspections can be regarded as, at best, setting a floor for the inspection process; but compliance is not necessarily fully responsive to the need to protect against major threats. The nation's willingness to accept hazards to the public and to the environment has declined sharply since the time when the OCSLA was last amended. Even the most rigorous compliance with its literal provisions is not likely to be considered good stewardship or management in the event of an accident, especially if it becomes evident that practical and reasonably available preventive measures were not required by regulators or taken by operators. In that light, the question of "how safe is safe enough" becomes important. Appendix F discusses modern safety goals and standards, and approaches used to meet those goals.

4

ALTERNATIVE APPROACHES TO SAFETY INSPECTION

Current Minerals Management Service (MMS) inspection practice for offshore platforms was described in Chapter 2. In this chapter, five alternative approaches to inspection are described. Three of them essentially are modified versions of the present system that could improve aspects of the inspection program from the standpoint of either safety or program management. They are

- alternative 1; increased inspection onsite by MMS;
- alternative 2; inspection of a sampling of potential incidents of non-compliance (PINCs) during annual inspections, and increased spot inspections; and
- alternative 3; annual inspection of a sampling of facilities, and increased spot inspections.

The other two alternative approaches are true alternatives, in that they involve fundamental departures from current practice. They are

- alternative 4; third-party inspection with government audit, and
- alternative 5; self-inspection.

An important point is that those alternatives that propose some form of reduced onsite inspection relate to production facilities only. The potential for accidents is higher in drilling operations than in production operations, which are more stable and predictable. Most continuous spills and some of the most serious accidents have resulted from drilling. Redundant safety devices and backup systems have been incorporated into the production process to a degree that is not feasible for drilling operations. Furthermore, safety of drilling operations is more directly affected by human behavior and its sometimes unpredictable sub-seabed conditions. There are far fewer drilling facilities than production platforms on the outer continental shelf (OCS), nevertheless, because of the large crews needed for drilling operations, during periods of development in the Gulf of Mexico the total number of personnel associated with mobile offshore drilling units (MODUs) and platform drilling equipment may be slightly higher than the number of personnel on production platforms (for example 10,551 drilling to 10,385 production, according to Coast Guard data[1]). Therefore, the committee believes that the present program of annual and spot inspections of drilling and workover activities should be continued. This is not to say that the inspection program enhancements discussed in this report should not be applied to inspection of drilling operations–they most assuredly should!

---

[1]This figure represents an average, over the period 1980-1986. The number of drilling operations in the U.S. OCS dropped drastically (by 50 percent) in 1986 because of the effect of low crude oil prices, and the OCS drilling population was correspondingly lower.

The next chapter presents several key considerations that the committee took into account in evaluating the alternative OCS inspection programs for efficacy and suitability. Chapter 6 discusses the results of the committee's assessment of both the present MMS program and the alternative approaches in light of these considerations.

MODIFICATIONS OF PRESENT PRACTICES

Alternative 1: Increased Inspection Onsite by MMS

This alternative involves modifying the present program by increasing MMS announced inspection activity over current levels. MMS inspectors would witness all mandatory tests and safety drills, i.e., all prescribed weekly, monthly, quarterly, and semiannual tests, in addition to witnessing tests of all devices once a year (the current goal) while continuing to perform occasional spot inspections. The primary objective would be to reduce the likelihood of incomplete or inaccurate test data in the first instance, since from a compliance viewpoint spot inspections could be targeted at facilities believed to be deficient in recordkeeping. An increase in scheduled witnessing of tests would add substantially to the manpower requirements of the MMS inspection program.

Alternative 2: Inspection of a Sampling of PINCs During Annual Inspections, and Increased Spot Inspections

Under this alternative, MMS would

- introduce a formal sampling program for conducting annual inspections to confirm that operators had accomplished prescribed tests properly;
- initiate a program of formal safety analyses; and
- perform a larger number of targeted spot inspections based on the results of the annual inspections and the safety analyses.

The immediate consequence of this alternative is to reduce the amount of time MMS inspectors spend onsite on production facilities to carry out the annual inspection and to compensate for the reduction with a program of more sophisticated safety analysis supporting more (and better targeted) spot inspections.[2]

The program contemplates also that steps would be taken to establish more clear-cut operator accountability for facility safety by requiring operators to

- record all routine tests of safety equipment by operating personnel, as well as those conducted at scheduled intervals prescribed by MMS (test results should be recorded in a format acceptable to MMS);
- establish a formal program for the operator's supervisors to follow up on their personnel's adherence to test schedules, to review test results, and to report the audit results periodically to MMS; and
- conduct internal (by management) reviews to ensure the performance of prescribed tests–performance of the prescribed internal review would be confirmed by MMS inspectors during the annual inspection, using the MMS-developed sampling plan.

---

[2]The committee believes that a shift to this alternative could be undertaken with acceptable risk while the essential task of enhancing the usefulness of the current MMS database is in progress.

Illustrative Scenario

The following is an illustrative scenario: several days prior to the annual inspection, the MMS would request that the operator send to the local MMS district office a certified copy of specified test and drill records for the platform. This might include, for example, all records for any particular item or all records covering a specified time span since the last annual inspection. The operator would have 48 hours after notification to submit the records. The MMS inspectors would review the records in their offices for adequacy and for possible trends and would plan for the best use of MMS offshore inspection time. The efficiency of this activity would be enhanced by using a sampling program in which limited numbers of like devices would be tested based on the operator's past safety history, the test result data accumulated by the operator, and the number of such devices installed on the facility. A facility-specific sampling plan could be developed based on the operator's inspection records, which would allow moving from the present program that emphasizes monitoring of tests to a program emphasizing verification of the operator's inspection records. Appendix G provides detailed commentary on the methods and feasibility of sampling.

Greater Operator Accountability

The present inspection program emphasizes an MMS role of detecting discrepancies affecting safety by annually witnessing a test of every safety device. Alternative 2 is intended to emphasize the responsibility of operators for making good safety practices an integral element of production operations. That is to say, this alternative would bring home to operating personnel and their immediate supervisors that they have full responsibility for the day-to-day safety of a facility; operator management would be accountable to closely monitor safety practices on its facilities; and MMS would use an increased number of spot inspections to verify adherence to good safety practices, rather than expending its resources in witnessing the repetitive mechanical testing of safety equipment.

The transition from what has become a program of monitoring of testing to what would be a program based on verifying the operators' inspection results will require thorough retraining of MMS inspectors, operator management, and particularly, operating personnel. Operator training programs will have to emphasize the need for close rapport between operating personnel and management on safety-related matters.

One difficult problem for the operator will be that, in order to improve the usefulness of the inspection data base, it will be necessary to record discrepancies even if they are corrected immediately. Periodically, the onsite supervisor will have to review tests and inspections to ensure that operating personnel have performed tests and inspections correctly. These reviews will have to be noted in the facility records and signed by the onsite supervisor. Also, operator management will have to be required to formally review the onsite supervisors' performance and record the results. With this type of system in place, MMS inspectors would be able to screen the facility records rapidly during spot inspections and identify the problems being encountered. Similarly, the follow-up being performed by the onsite supervisor and the operator's management could be tracked through the facility's records.

Safety Analyses

Analysis of test results, past violations and events (accidents), and other safety-related data would be performed by the MMS as a distinct inspection activity (see Chapter 7 for elaboration). In-depth analyses could increase the probability of safe operations in several ways. For example, during the committee's visit to a Pacific Coast platform, an MMS inspector took time to verify that repairs had been made to six relief valves that MMS had rejected on the previous inspection. The

MMS inspector stated that this number of discrepancies involving relief valves was common in the region. In the view of the committee, this was an excessively high failure rate for such a critical device, particularly in light of the frequency of MMS inspections of the facility. Ongoing safety analyses of inspection data might have flagged the possibility of a failure trend requiring remedial action.

Where potentially important findings had been made by MMS inspectors performing these analyses, the results would be provided to MMS management and to operators along with background and recommendations for appropriate action. Such actions could take the form of changes in procedures or specifications, modifications in hardware design, or reallocations of test or inspection effort.

Enhanced Spot Inspections

By reducing the time devoted to annual inspections, more spot inspections could be carried out. Like the annual inspections, these spot inspections also would rely on sampling techniques for the selection of PINCs to be checked; similar techniques could be used for selecting the facilities to be visited. More spot inspections would offer the dual benefit of increasing the MMS presence on all OCS facilities, and making operators more continually aware of their safety responsibilities and performance.

Based on the results of safety analyses, MMS could use the increased spot inspections to exert a concerted effort to address identified problem areas. PINC sampling technique could thus be tied to safety analysis methodology to yield highly targeted spot inspections.

Alternative 3: Annual Inspection of a Sampling of Facilities, and Increased Spot Inspections

With an adequate data base reflecting the safety history of individual facilities it would be possible to develop a sampling plan to conduct annual inspections only at selected facilities. The sampling plan would require data for ranking each facility on the basis of its safety history, the safety history of the operator, the location of the facility, and the safety history of like facilities.

An alternative built on this approach would allow MMS inspectors to focus their annual inspections on those facilities identified as being more susceptible to having safety problems. It would reduce the amount of time spent in onsite inspection by MMS inspectors for annual inspections of production facilities by reducing the number of annual inspections performed each year. As in alternative 2, the resources thus freed would have to be used to increase spot inspections and to institute a safety analysis program. An element of this alternative would be that each facility be subject to at least one onsite spot inspection each year.

The key to an effective selected-facility inspection program is the development of a valid sampling plan. In the case of a selected-facility inspection program the sampling plan would be particularly critical since the annual inspection of some facilities would be foregone. In order to develop the sampling plan, more extensive data will be needed regarding the operators and facilities than currently is available. Adoption of this alternative will have to be preceded by an intensified data-gathering period in which all facilities would continue to be subject to annual inspections. (Introduction of the sampling program for PINCs would not interfere with the data collection effort). The annual inspections of the selected facilities by the MMS could extend either to the witnessing of 100 percent of the tests of safety devices, as is currently required, or more probably to an inspection limited to witnessing a sampling of PINCs. In the committee's view, the use of a sampling plan for selecting the items to be checked during selected-facility annual inspections would not result in an unacceptable level of safety.

The credibility of the selected-facility annual inspection program would require that each facility be visited at least once a year for spot inspection. Spot inspections could be as abbreviated as a simple "walk-around." The findings in spot inspections, event (accident) investigations, as well as those in the annual inspections of the selected facilities, would be used to refine the selected-facility and PINC sampling plans and as a basis for the issuance of safety alerts.

DEPARTURES FROM PRESENT INSPECTION PRACTICE

Alternative 4: Third-party Inspection with Government Audit

Under this alternative, MMS would mandate that operators contract with independent, certified third-party inspectors to perform the annual and spot inspections now performed by MMS personnel. MMS would certify the third-party inspectors on the basis of specific certification criteria, and periodically would conduct audits of the inspections conducted by the third parties to establish that they were performed properly. MMS also would monitor and verify the continuing qualifications of the third-party inspectors. MMS would specify the documentation required to be completed and submitted by the third-party inspectors to record that an acceptable annual or spot inspection had been performed and would institute a follow-up program to see that the prescribed inspections were performed for each platform. This inspection program alternative is similar to the program currently employed in the United Kingdom sector of the North Sea.

One element of this alternative that would distinguish it from current practice is that several considerations (the large number of variations in local circumstances, the need to establish the private inspection organization's responsibilities with certainty, and the removal of MMS personnel from day-to-day activity on the platforms) would require the operators and/or the private inspection organizations to formulate a safety plan for MMS approval that describes the inspection processes to be followed. The mere process of preparing a safety plan (which would require a detailed analysis of the operation) could yield safety benefits.

To implement this alternative, the MMS would have to establish criteria for certifying third-party inspectors and guidelines for preparing safety plans and procedure manuals to be submitted by operators for MMS approval. The manuals would specify how the required inspections would be performed. Two variations are readily evident:

1. The operator could chose to perform all mandated tests and drills using his own personnel and use a certified third-party inspector to audit the performance of the mandated periodic tests and drills and to perform the annual inspection; or
2. the operator could elect to have a certified third-party inspector perform all the mandated tests and drills as well as the annual inspection.

Whatever variation is selected, an operator would not be permitted to perform the annual inspection of his own facility.

It would be essential that the MMS establish a formal procedure for certifying the qualifications of third-party inspection services and for auditing their performance. The procedure would have to ensure that conflict-of-interest situations are avoided. It also would have to provide for continuous monitoring to ensure that the third-party inspector's technical capabilities and personnel resources for carrying out a safety inspection program are maintained in the face of turnovers of either staff or management.

By conducting periodic audits of the performance of the certified third-party inspectors, and by requiring them to be certified at regular intervals, the MMS would be able to measure the effectiveness of the program and ensure that adequate inspections are carried out. Third-party inspectors would be subject to losing their certification if they did not perform effectively.

A variation of this alternative would be for the MMS to hire the third-party inspection services directly. This alternative would involve less of a possibility of a conflict of interest between the third-party inspector and the operator. However, it would not serve as effectively to overcome the unwanted perception that safety on OCS facilities is primarily the responsibility of the MMS (through enforcement of compliance with PINCs) rather than the operator (through safety management.)

If the MMS were to hire a third-party inspection service directly, the change would amount to no more than replacing government inspectors with private-sector inspectors. Thus, there would be no substantive improvement in the efficiency of the performance of the inspection program, although it is arguable that more flexible utilization of personnel might be achievable (with concomitant cost savings.)

Alternative 5: Self-inspection

The MMS would require that the operators inspect themselves, i.e., that they conduct all tests and inspections without MMS witnessing. In its ultimate form, the alternative would be similar to the program recently announced by the U.S. Coast Guard, in which operators report the results of their annual inspections on a standardized form, with the Coast Guard reviewing these reports and verifying them with spot inspections. This alternative obviously would produce substantial savings in manpower and budgetary resources, some of which the MMS could utilize to conduct more spot inspections and needed in-depth safety analyses of OCS facilities and the safety programs of the individual operators. Even with more spot inspections, this alternative could well serve to reduce MMS personnel day-to-day contact with OCS operations to unacceptable levels.

5

CONSIDERATIONS IN EVALUATING OCS INSPECTION ALTERNATIVES

Having identified five possible alternative approaches to the present inspection system, the committee sought to develop a process for evaluating the relative merits of the alternatives (as well as the present practice). Eight considerations were identified that an inspection program ought to address and that should be considered in evaluating the alternatives:

1. Does the alternative being considered promote industry safety awareness?
2. Does it help to maintain public confidence in the safety of OCS operations?
3. Does it use inspection resources efficiently?
4. What is the impact on the qualification and training of the inspector force?
5. Does the program provide for identification of safety trends and warnings?
6. Does it promote safety performance accountability?
7. Is it adaptable to changing circumstances?
8. Are there valid precedents?

These questions are elaborated upon below and the alternative programs evaluated in Chapter 6.

PROMOTION OF INDUSTRY SAFETY AWARENESS

Several points are relevant here. First, the main objective of any inspection program is to promote industry safety awareness at all levels, including supervision and management. The program should provide timely feedback to operating and supervisory personnel as well as to the management personnel who have control of resources.

Another consideration is whether there are sufficiently large penalties and incentives built into the system to ensure timely attention and action by managers in control of resources. Especially for alternatives 4 and 5 (third-party and self-inspection), it is relevant to ask whether the inspection system and the associated data and feedback process can support civil[1] and, if necessary, criminal penalties for persistent nonperformance and/or non-compliance with regulations, recommendations, or requirements emerging from inspections.

There should be appropriate, timely, and effective follow-up reviews and/or inspections, and a means to communicate to operators the degree of criticality of observed deficiencies, to ensure an

[1] Currently the civil penalty provisions of the OCSLA are not an effective enforcement tool. Court decisions require MMS to afford an operator time to correct a deficiency before imposing a civil penalty. In effect, the operator never is subject to civil penalties.

appropriate priority to corrective action, and to determine and support enforcement actions and criminal penalties when these are required.

The goal of enhancing industry awareness of safety factors implies a requirement for an effective safety evaluation methodology and for a significantly increased level of enforcement and follow-up when crucial deficiencies are observed or reported. It is the committee's view that meeting the minimum requirements of the Outer Continental Shelf Lands Act (OCSLA) regulations does not ensure the high levels of safety that should be demanded of large industrial operations involving potentially hazardous or polluting materials. Effective ways to accomplish these objectives can be built into any safety inspection program. It is easier, however, to achieve the goal with a program that stresses accountability rather than one that stresses compliance.

MAINTAINING PUBLIC CONFIDENCE

In general, MMS must ensure that the actual safety performance of OCS operators and the public perception of it engenders a consensus that MMS is effectively overseeing public interests on the OCS. Several different areas of concern not now addressed by MMS must be considered when evaluating whether an alternative inspection system would contribute to public confidence concerning the safety of OCS operations.

View of Congress

The MMS traditionally has focused its efforts on literal compliance with the inspection function described in the OCSLA, i.e., to conduct annual and spot inspections.[2] In the event of a serious accident on the OCS, such as the one that destroyed the Occidental Oil platform Piper Alpha in the North Sea in July 1988, it is likely that the Congress would conclude that literal compliance with the OCSLA reflected only weak credit on MMS's stewardship, if other risk-reducing activities of the general kind taken by federal agencies regulating other risk-intensive operations had not been taken. Most of the alternative inspection programs described in Chapter 4 contemplate the addition of modest planning and risk-analysis activities that could be undertaken within current budgetary levels. These activities are not being carried out at present.

It is likely that such analytical efforts would lead to changes in the priorities, processes, and performance of some of the inspections. The adoption of modern system safety techniques for inspection and enforcement probably would provide a more meaningful way to measure compliance with the intent of the OCSLA (see Appendix C). Apart from the direct benefits such improvements would bring, they would also help in the event of a serious accident to deflect criticism that MMS has not been sufficiently aggressive in promoting safety.

Media and Public Opinion

The media have a powerful impact on public opinion. Currently there is intense interest in the media as to how well U.S. inspection and enforcement practices help prevent accidents. This interest has grown out of such events as Challenger, Bhopal, Chernobyl, a spate of aircraft structural failures, breakdowns in government-mandated airport security measures, and the *Exxon Valdez* oil spill.

---

[2]As noted in Chapter 3, and discussed in Appendix G, the MMS's view of its safety role is focused on Section 1348 of the OCSLA (43 U.S.C. 1331-1356) and may not reflect the broader overall policy mandate of the OCSLA.

Fine-tuning of inspection procedures, personnel practices, and organizational roles in the ways discussed in the five alternatives is desirable and appropriate. But these actions would not seem to be sufficient to respond to present public expectations regarding safety and environmental protection. An explicit and defensible rationale for establishing priorities and standards and providing at least a general basis to reply to the question "How safe is safe enough?" is urgently needed to give MMS's and the industry's safety consciousness credibility.

While the perception of OCS operations held by the local community is largely created by the media, it also is a function of the interaction of the operator with local government. These are matters beyond the direct control of the MMS.

Comparability With Other Federal and State Agency Practices

Present MMS inspection practices appear to be more rigorous than the counterpart state practices. However, other federal agencies have more extensively documented rationales underpinning their practices in inspection, testing, and enforcement to implement the intent of legislation analogous to the OCSLA in their areas of responsibility. The Environmental Protection Agency, the Food and Drug Administration, the Occupational Safety and Health Administration, and the Nuclear Regulatory Agency are examples.

EFFICIENT USE OF RESOURCES

An important consideration in assessing any inspection scheme is how efficiently it utilizes resources–personnel, time, and money. Both government and operator resources are at issue.

Government Resources

An alternative inspection program that reduces MMS offshore inspection time will reduce government costs. To the extent that these savings can be realized without sacrificing safety, there will be an overall improvement in the benefits of inspections; the benefit will be even greater if the alternative better promotes safety.

Any of the alternatives, as well as the present program, can be improved to minimize the amount of paperwork done offshore and to make more efficient use of helicopter time. At the facility, hands-on inspection that can be done only onsite needs to be emphasized over paper reviews that can be carried out onshore. An improved system should provide incentives to organize paperwork to maximize the effectiveness of onsite inspection time.

To some extent, scheduling is independent of the specific program alternative selected, but some alternatives may be better adapted to accomplish these objectives. The scheduling process should include provisions for moving people from one office to another when peak loads occur for inspection, reinspection, or follow-up.

Any near-term change in the MMS inspection system should emphasize greater productivity and more efficient use of current levels of manpower and budget. If the decision is made to adopt an inspection system with enhanced data collection and safety analysis capabilities, more efficient use of available inspection time will be necessary to offset the increased cost of enhancing these capabilities. Greater efficiency in use of inspection time could be obtained through the use of sampling techniques, for example, with no decrease in safety and with lower cost.

Opportunities for leveraging MMS inspections should be sought. For example, a system of thorough, well-documented inspections by an operator, contractor, or other qualified third party could be structured to be acceptable as annual inspections without detracting from safety.

Operator Resources

Operators on the OCS perform thousands of inspections and tests each year. Except for the unscheduled, unannounced checks, MMS inspections ideally should make use of the operator's planned schedule for tests and inspections whenever there is a history of good compliance and few deviations from reported conditions or in test results.

An alternative inspection program ought not to impose large increases in the operator's cost of compliance, viewed in terms of the industry average of costs. (That is, only operators with a poor safety program likely would incur high costs in bringing their practices up to standard.) A reasonable increase in recordkeeping costs can be absorbed, but the largest costs ought to be front-end setup costs.

Societal Costs and Benefits

Modest increases in the government costs of OCS safety inspection and enforcement are not likely to provoke public objection. The budgets involved are small and are distributed nationally. Furthermore, public surveys and voting patterns indicate a willingness on the part of the public to pay for increased safety and environmental protection activities, provided those activities are seen as effective, i.e., few or no major events occur.

Major events (blowouts, spills, fires, etc.) have large direct and indirect costs to the public as well as to the owners and operators. These costs appear in the form of federal and state taxes, insurance fees, public medical expenses for families of people killed or injured, the intangible human emotional costs of losses of life and property, and the costs of damage or destruction of environmental resources and cleanup. Any inspection system that increases safety produces cost savings in these areas.

QUALIFICATION AND TRAINING OF INSPECTOR FORCE

A change in the inspection program could have an impact on the basic qualifications and training of the MMS inspector force. If the change is significant, there will be an associated cost for retraining and possibly for a complete restructuring of the staff.

An inspection program should involve uniform inspection processes and performance standards. However, as was discussed in Chapter 3, rigid adherence to checklists in order to achieve uniformity can lead to a compliance mentality, to the detriment of overall safety. An inspection program requires a balance between specific criteria for equipment tests and procedures on the one hand, and a necessarily subjective analysis of safety attitudes, training, and management commitment on the other. The greater the emphasis on compliance with rigid standards, the easier it is to train the work force. However, the real objective is greater safety, not facilitating training; and the subjective elements will have to be emphasized strongly if inspectors are to inspect for adherence to performance standards.

IDENTIFICATION OF SAFETY TRENDS AND WARNINGS

An inspection program should provide the means to identify safety trends and spot warning signals of impending problems. This implies that the inspection system must include data collection, reporting, and safety analysis systems to provide a high probability of detecting and diagnosing conditions or trends that are likely precursors to accidents. The data systems should provide a basis for focused follow-up and enforcement actions when these are indicated.

Data systems encompass the safety management systems and operating procedures of operators, including associated training manuals and curricula, frequency of training and measures of efficiency, and procedures for dealing with abnormal conditions, including major and minor accidents.

The reporting and review elements of an inspection system should include ways to measure the adequacy of coverage, frequency, and probable error or "miss" rates of the operator's inspections. There should be sufficient definition of these factors to provide a basis for determining whether acceptable levels of safety can be expected from reliance on inspection data of operators who have excellent safety records and a demonstrated corporate commitment to safety.

Any of the alternative inspection programs described in Chapter 4 could be designed to permit collection and analysis of such data. However, the various alternatives differ in the extent to which MMS inspectors would witness and verify the accuracy of data collected. As a result, they differ in the amount of resources that would be released for safety analyses. The greater the extent to which MMS personnel witness data collection, the more reliable the data are likely to be; correspondingly, the less available will be funds and personnel for analytical activities and spot inspections (unless the budget is increased). The most effective inspection program in this regard will be one that reduces MMS witnessing of data collection without affecting data reliability, while simultaneously freeing more resources for analytical activities and spot inspections.

SAFETY PERFORMANCE ACCOUNTABILITY

The candidate inspection systems encompass a variety of ways to monitor performance of MMS personnel, operators, and the inspection system itself. The data collection and safety analysis effort required to produce safety trend information will provide much of the basic data needed for monitoring of operator performance. Any of the alternative programs can be designed to incorporate a system of accountability for individual MMS inspectors, supervisors, and district offices. The inspection system must provide measures of the effectiveness of each element of the inspection process, and of the effectiveness of the overall process in supporting the ability of management (both MMS and operator) to observe and correct deficiencies.

Measures of MMS managerial effectiveness should allow for the different degrees of complexity of facilities in different areas and differing travel requirements. While easily measured statistics such as the number of inspections and INC/PINC (incidents of non-compliance/potential incidents of non-compliance) ratios can be useful in gauging productivity, the real goal should be to hold MMS personnel responsible for the degree to which they bring out good safety procedures and attitudes on the part of the operators. This can be difficult to measure objectively. Safety records of the various operators in a given geographic area and steps taken by MMS personnel to encourage improved safety are two categories of data that might aid in this determination. The extent to which the various alternative programs facilitate the collection and analysis of such data is thus a major point to consider in selecting an alternative.

With any alternative being considered, provision must be made for avoiding conflicts of interest. The potential for conflicts of interest is especially pertinent in alternative 4, third-party inspection. The operator contracting for third-party inspection services may be tempted to look for a "responsive" contractor; the contractor or his employees in turn may want to ensure continued employment and thus be tempted to be responsive. With the MMS offshore presence being greatly reduced, it would be imperative to guard against the potential for these circumstances to denigrate safety.

ADAPTABILITY OF THE INSPECTION SYSTEM TO CHANGING CIRCUMSTANCES

An inspection system should be flexible and able to evolve in the face of changing conditions, public perceptions, and regulatory requirements. This is particularly important with regard to the following factors:

- Technology: the technology of structures, exploration, and production is undergoing continual refinement. The technology of safety assessment and control is also evolving and at an even faster rate. An inspection system should include a technology update/watch and a procedure to incorporate important developments in safety technology into existing operations and new facilities.
- Inspection methods: the nature and coverage of the inspection should be adaptable to fit changing needs. It should take advantage of developments in safety/risk analysis. For example, two important near-term developments in safety technology have so far found little application in the MMS inspection program: (1) the increased use of systematic risk assessment to set priorities and key target items for inspections, and (2) the development of improved sensor and computer technology to extend the effectiveness and efficiency of personnel.
- Aging facilities: as platforms approach the end of their useful lives, the spectrum of hazards to be addressed by the inspections should be expanded. A continual process of review and updating of inspection targets and priorities is required, together with a structured approach to defining the end of a facility's useful life. Acceptable measurements have to be defined to support extension of the service life of old facilities.

PRECEDENTS

The consideration here is whether the inspection procedure follows a generally accepted pattern and whether it incorporates a process for maintaining MMS awareness of relevant experience in other regions, other industries, and other countries.

- Related oilfield and platform experience: surveys of related experience in oilfields and on OCS platforms in the past (including records of events in state waters) should be conducted. Periodically these surveys should be updated and analyzed, especially to identify those circumstances that contributed to major breakdowns or failures; the results should be made available for use in selecting and constructing better systems.
- Related experience in other industries: current practices and related experience in other industries in which a substantial risk of public safety hazards or an environmental pollution potential may exist should be reviewed continually to keep abreast of technological developments having possible application to OCS activities. There should be a mechanism for assessing the impact of changes in public expectations as to safety and environmental pollution.
- Related experience in other countries: awareness of current practices and related experience—especially accident experience—in other countries can provide important additions to the U.S. data base. Experience, not only in the oil and gas industry but also in other industrial activities where potential hazards to public safety and/or the environment are present, can be useful.

GENERAL CONSIDERATIONS

Underlying the eight considerations discussed above is a fundamental requirement that any acceptable MMS inspection program must satisfy: it must contribute toward the overall goal of safely operating facilities on the OCS. This basic goal includes

1. the safety of personnel–that is, that the selected program encompasses training, procedural, and operational guidelines that will promote operational safety and reduce lost-time injuries and fatalities;
2. protection of property and facilities;
3. environmental protection–that is, the program incorporates equipment monitoring and operating practices that provide acceptable margins of safety against spills, blowouts, and fires or explosions; and
4. preservation and conservation of energy resources.

Primarily, the MMS inspection program must be responsive to its congressional mandate. The OCSLA requires an annual inspection of every facility and a program of spot inspections. This requirement can be interpreted as a congressional determination that a comprehensive annual inspection by MMS is essential to achieve important public and national goals including overall safety. However, the OCSLA is not explicit about what constitutes an inspection or who performs it, as long as the MMS inspection program ensures that those goals are being achieved. Thus it would appear that MMS is free to implement measures it believes would be more effective than current practices in achieving its safety goals (based on MMS's interpretation that the law requires its inspectors to witness testing of every safety device annually). If MMS concludes that existing law impedes the adoption of measures it believes would enhance safety on the OCS, clarifying legislation could be sought.

6

ASSESSMENT OF PRESENT PROGRAM AND ALTERNATIVES

The committee assessed the present Minerals Management Service (MMS) inspection program as well as the five alternative programs described in Chapter 4. The assessment was made on the basis of the considerations described in Chapter 5.

PRESENT PROGRAM

The present program of regulation and inspection was initiated in the aftermath of several major offshore accidents, involving both loss of life and pollution of the environment, which occurred in the 1968-1971 period. The program was developed predominantly on the basis of industry recommendations. Given the urgent need for action at the time, this approach probably was the only practical solution. Since that time, after nearly 20 years of regulation and monitoring for compliance by USGS, and later by MMS inspectors, there clearly has been improvement in operational safety and in industry's attitude toward safety. Nevertheless, for the reasons outlined in Chapter 1 (see "Summary") and because of advancements in safety technology, a fresh look at the effectiveness of the present program is indicated.

Responsiveness in Meeting Present Safety Needs: Adaptability to Changing Circumstances

The present MMS inspection program is responsive to the express terms of the Outer Continental Shelf Lands Act (OCSLA)–i.e., (1) annual announced inspections are conducted of essentially all OCS facilities subject to environmental and safety regulations, and (2) unannounced spot inspections are conducted of a reasonable proportion of the facilities (although manpower constraints in the Gulf of Mexico have resulted in fewer spot inspections than programmed). More important, however, is whether the substantive safety objectives of the OCSLA are being achieved. These objectives–occupational (personnel) safety, environmental protection, resource conservation, and protection of property (primarily OCS facilities)–are stated only in general terms in the statute and their purview is dependent on specific guidance promulgated by MMS.

This guidance currently is reflected in the national outer continental shelf (OCS) orders and the potential incidents of non-compliance (PINC) list described in Chapter 2. On the one hand, the committee's study led it to question the adequacy of that guidance in addressing the full spectrum of safety issues. On the other hand, the committee has come to believe that a number of PINCs require inspector activity that may be superfluous, given the essential checks that the operator must conduct to maintain his operation at an acceptable level of productivity.

To determine whether the PINC list addresses items that historically have contributed to accidents, the committee reviewed the MMS "Events File," which contains brief summaries of accidents (see Figure 6-1 for a typical listing) for cases in the period 1982-1986 involving deaths in the Gulf of Mexico.

The committee found that there were 63 deaths and 31 injuries caused by the 54 events in the period. (Table 2-12 presented an overview of these findings.) Each event was classified by basic root cause, to the extent this could be derived from the file, and the causes were listed in order of number of events. An attempt was then made to identify any PINC violations that could have contributed to the cause of the event. Six of the events (11 percent) had either an unknown cause or one completely unrelated to MMS responsibility. Another 45 of the events (80 percent) were covered (if at all) only by the "all-purpose" PINC number G-400 ("Is each operation performed in a safe and workmanlike manner and are the necessary precautions taken to prevent accidents?"). Of the remaining 9 percent, two electrical events possibly could have been caused by violations of G-300 ("Is electrical equipment installed, protected, and maintained in accordance with the National Electric Code?"); two events caused by gas being vented during well workovers could be attributed to a violation of D-408 (condition of valves, pipe and fittings); and one event could have been caused by violation of one of several welding PINCs. *None of the events was caused by or related to failure of a device whose testing is currently being witnessed by the MMS.*

While this result could be seen as proof of the effectiveness of the MMS program, it is the committee's perception that even though the program undoubtedly has advanced the safety of OCS operations the foregoing result indicates that it is not sufficiently broad in scope. The following findings give further weight to this perception.

The committee reviewed the events file for 1982 for deaths, injuries, fires, and pollution in Gulf of Mexico production operations. It was necessary to go through the listings for all 263 events to develop Tables 6-1 through 6-6, which group the events by cause.[1] Only 42 (16 percent) of the events could have been caused by a specific incident of non-compliance (INC) other than G-400. Of these, 19 (7 percent of total events) were associated with small oil spills from overflow of drip pans and sumps (E-103 and E-100). Seven (3 percent of the total) potentially were associated with electrical system INCs (G-300) and five (2 percent) with welding operation PINCs (G-206). Four events could have been caused by violation of PINC P-301 ("Is engine exhaust insulated and piped away from the fuel source?"). Six events (2 percent) potentially were due to violations pertaining to shutdown devices. Of these six, two events involved devices that were bypassed at the time of the event and two involved devices that were required but not installed; only PINCs P-167 and possibly P-700 (one event each) potentially involved safety devices that failed to operate (see Table 6-6). (It also is possible that some of the sump overflows could have been due to failure of a sump high-level shutdown, but this is not apparent from the data.) In any case, it is clear that very few of the events were caused by the failure of a safety device to activate.

The rarity of events caused by failure of safety devices does not prove that the regulations requiring the installation and testing of these devices are unwarranted. It is likely that these requirements are preventing events that otherwise would have occurred. *However, the data do indicate that further reduction of events cannot be accomplished by mandating more stringent testing, inspection, or quality control procedures with regard to the already mandated safety devices.*[2]

---

[1]The total of events cited in Tables 6-1 through 6-4 is larger than 263 due to attribution of multiple causes to events in some cases.

[2]One of the committee members who specializes in risk management, Dr. Edwin L. Zebroski, has studied the Piper Alpha platform disaster. He notes that the PINC inspection process, even using the most liberal interpretation of item G-400, would have picked up few if any, of the many deficiencies in operation, staffing, and maintenance practices that contributed to that catastrophic event.

FATALITIES 1982-1986 – OCS EVENTS FILE

File Number --
Table List -- In Accident Report
Area and Block -------
OCS Area ------------- GOM
District Office -------
Lease Number ---------
Water Depth ---------- 0240'
Distance from Shore -- 079 Mi   *Structure ---*
Operator -------------

Type of Operation
Production
Type of Facility
Type of Production Facility
Steel Tower

Type of Event(s)
Fire
Injury
Fatality
List --

Cause of Event
Other vessels and tanks (incl. surge and prod. tanks)
other (specify) – Pressure buildup in tank
Source of ignition
Hot exhaust
List --

Effects of Event
Number of men injured ---------------- 3
Number of fatalities -------------------- 1

General Remarks
A platform well service employee was fatally burned when he was sprayed with diesel fuel as he removed a plug from the top of a diesel storage tank to check the fuel level. Apparently, filling operations by the [vessel's] pumps were not completely shut down before the deceased had replaced the tank vent plug, which caused a pressure buildup in the tank. Ignition occurred immediately and was most likely caused by the generator engine's exhaust directly overhead.

Date of event:

FIGURE 6-1 Typical listing of accidents in the MMS Events File.

SOURCE: MMS, OCS Events File. [Information that identifies the specific platform has been deleted.]

TABLE 6-1 Evaluation of Event Listing for Accidents Causing Deaths in the Gulf of Mexico, 1982

Description of cause	Number of events	Number of deaths	Potential PINC violations
1. Falling from height	2	2	G-400
2. Opening a pressurized system for maintenance	2	2	G-400
3. Drowning	2	2	G-400
4. Helicopter accidents	1	2	NONE
5. Unknown	1	2	UNKNOWN
6. Design violations	1	1	P-653[a]
7. Handling heavy loads	1	1	G-400
8. Illness/heart attack/overexertion	1	1	NONE
Totals	11	13	

[a]This may have been a system operation, in which case it would have been listed under cause #2.

TABLE 6-2 Evaluation of Event Listing for Accidents Causing Injuries in the Gulf of Mexico, 1982

Description of cause	Number of events	Number of injuries	Potential PINC violations
1. Falling from height	36	37	G-400
2. Handling heavy loads (including crane accidents)	35	35	G-400
3. Loss of footing/walking into objects	25	25	G-400
4. Improper use of tools or equipment	17	17	G-400
5. Opening pressurized equipment	11	15	G-400
6. Engine/compressor/turbine maintenance and operation	6	11	G-400
7. Boat accidents	5	9	G-400
8. Sandblasting operations	5	5	G-400
9. Illness/heart attack	4	4	NONE
10. Welding and cutting operations	4	5	G-206, G-400
11. Walkway failures	3	3	NONE
12. Drain and sump systems	1	4	G-400
13. Design violations	1	3	P-653
14. Helicopter accidents	1	2	NONE
15. Electrical shorting	1	1	G-300
16. Diving operations	1	1	G-400
17. Unknown	1	1	UNKNOWN
Totals	157	178	

TABLE 6-3 Evaluation of Event Listing for Accidents Causing Pollution in the Gulf of Mexico, 1982

Description of cause	Number of events	Number of barrels	Potential PINC violations
1. Unknown	15	38	UNKNOWN
2. Drain and sump systems	13	34	E-103
3. Liquid discharged through vent	8	18	NONE
4. Handling heavy loads	5	16	G-400
5. Pipline leak/failures	5	15	E-100, E-104
6. Equipment failures	5	13	E-103, P-653
7. Drip pan design	3	13	E-100
8. Boat collisions	2	6	G-400
9. Safety devices bypasses	2	6	P-100
10. Poor operating procedures	1	48	G-400
11. Welding and cutting operations	1	2	G-206
12. Opening pressurized system	1	2	G-400
13. Electrical shorting	1	1	G-300
14. Improper tool or equipment use	1	1	G-400
15. Control component failure	1	1	P-167
Totals	64	214	

TABLE 6-4 Evaluation of Event Listing for Accidents Causing Fires in the Gulf of Mexico, 1982

Description of cause	Number of events	Number injured	Potential PINC violations
1. Engine/compressor/turbine maintenance and operation	13	10	P-301, G-400
2. Unknown	8	1	UNKNOWN
3. Welding and cutting operations	7	1	G-206, G-400
4. Equipment failure	5	0	G-300
5. Electrical shorting	3	0	G-300
6. Opening pressurized system	2	3	G-400
7. Improper tool or equipment use	2	1	G-400
8. Poor operating procedures	2	0	G-400
9. Lightning	2	0	NONE
10. Drain and sump systems	1	4	G-400
11. Design violations	1	3	P-653
12. Sand blasting operations	1	1	G-400
13. Improper material storage	1	0	G-400
14. Liquid discharged through vent	1	0	NONE
15. Control component failure	1	0	P-700
Totals	50	24	

TABLE 6-5 Summary, 1982 Offshore Event Listing (Production)

	Number of events	Potential PINC violations
1. Handling heavy loads	41	G-400
2. Falling from height	37	G-400
3. Lost footing	25	G-400
4. Unknown	23	Unknown
5. Improper tool or equipment use	19	G-400
6. Opening pressurized systems	14	G-400
7. Drain and sump systems	14	G-400, E-103
8. Engine/compressor/turbine operations	14	P-301, G-400
9. Equipment failures	10	G-300, E-103, P-653
10. Welding and cutting operations	9	G-206, G-400
11. Liquid discharge from vent	9	NONE
12. Boat accidents	8	G-400
13. Electrical shorting	5	G-300
14. Sand blasting operations	5	G-400
15. Pipeline leak/failure	5	E-100, E-104
16. Illness/overexertion/heart attack	4	NONE, G-400
17. Drip pan design	3	E-100
18. Walkway failures	3	NONE
19. Poor operating procedures	3	G-400
20. Safety device bypass	2	P-100
21. Lightning	2	NONE
22. Control component failure	2	P-167, P-700
23. Drowning	2	G-400
24. Diving operations	1	G-400
25. Helicopter	1	NONE
26. Improper materials storage	1	G-400
27. Design violations	1	P-653
Totals	263	

It should be pointed out that the events files do not specify whether an INC caused or contributed to the magnitude of an event. Thus, the analysis of potential PINC violations in these six tables only indicates the INCs that potentially might have been related to the cause of the event; it does not indicate that an INC necessarily existed at the time of the event.

It was necessary for the committee to develop these data from the MMS's computer listings. Similar analyses of the events data by MMS were not available. Such analyses could, however, aid in the evolution of the inspection program.

The committee also analyzed 262 events in 1982 pertaining to drilling and workover operations, though not in as much detail as it did the events related to production operations. Table 6-7 summarizes these events by category of cause. No attempt was made to tie the cause of events to possible PINC violations. The drilling PINC list is concerned mainly with well design, operating procedures and tests, and equipment necessary to control wells to minimize the occurrence and consequences of well-control failures. Of the 1,606 wells drilled and worked over in 1982, only 8 incurred well-control failures. Devices that the MMS requires to be installed and

TABLE 6-6 1982 Event Listings - Summary by PINC Violation

PINC violation	Number of events
G-400	162
None	35
Unknown	24
E-103	14
G-300	7
G-206	5
E-100	5
P-301	4
P-653	2
P-100	2
E-104	1
P-167	1
P-700	1

TABLE 6-7 Summary of 1982 Offshore Event Listing (Drilling)

Cause of event	Number of events
Falling, handling heavy loads, etc.	209
Tank runover, ruptured hose	16
Electrical systems	9
Loss of well control	8
Pollution from mud circulating system	5
Welding	4
Opening a pressurized system	3
Equipment overpressure	1
Premature firing of perforating gun	1
Boat collision	2
Unknown	4
Totals	262

tested prevented fires in 4 of the 8 events. The extent of injuries or oil pollution in the course of the 8 blowouts–although clearly not major, based on collateral evidence regarding major accidents in 1982–cannot be determined from the events file, as this information is not reported for some of the events. It must be kept in mind that any well-control failure involves the threat of loss of life and a large spill.

In 1982 there were approximately 275 drilling or workover facilities and 2,900 production platforms in the Gulf of Mexico. The events file indicates that there were 0.95 events per drilling or workover facility and 0.09 events per production platform in that year. Therefore, for 1982, the number of events during drilling operations was 10 times that for production operations. (In periods of low drilling activity, such as 1986, the frequency of events was 3 to 1 for drilling operations compared to production operations.) As was noted earlier, the population of drilling personnel on the U.S. OCS in 1982 was at least as large as the population of production workers. As with production events, the majority of drilling events were related to human error and operating and maintenance procedures. The committee believes that a regular review of the events in drilling operations as they occur could lead to improved regulation and inspection techniques. For these and other reasons the committee believes that continued frequent onsite inspection of drilling rigs is important.

In the aggregate, the committee's analysis of the 1982 events file shows that the vast majority of events in both drilling and production operations resulted only in minor personnel injury, small flash fires, or small oil spills. These events, which stemmed from human error in operating and maintenance activity, suggest a lack of proper attitude and awareness toward safety on the part of workers, supervisors, and management. The current MMS regulations, with their emphasis on well control in drilling and workover operations, and process upsets in production operations, are designed to avoid major disasters. However, historically disasters frequently are preceded by low-level events caused by inattention and a poor attitude toward safety, deficiencies that can be identified by safety analysis.

Overall, the weaknesses of the present program are as follows:

- Full compliance with the testing requirements of the PINC list does not necessarily address essential safety elements comprehensively, especially those involved in the low-level events–many of which involve human error–that represent the majority of events (accidents).
- Operators may be motivated to concentrate their safety efforts on maintaining devices that they expect an MMS inspector will require them to test while following a PINC list, rather than on systematically addressing measures to improve the overall safety of their operations and to conduct them in a "safe and workmanlike manner."
- The emphasis in the PINCs on individual mechanical components inhibits a rigorous evaluation by MMS of the overall safety of the operation. In production operations, the PINC list and the ratio of INCs to PINCs to a large extent address the reliability of selected devices in operating within prescribed tolerances. MMS has not established that these data are, in themselves, sufficient indicators of safe operations. The PINCs themselves place almost total emphasis on mechanical tolerances and do not leave enough room for safety judgment.
- MMS's focus on incidents of non-compliance tends to obscure the fact that the responsibility for safety lies with the operator, and not with the MMS. In these circumstances, an evaluation of the commitment of the operator toward good safety practices, much less any erosion thereof, tends to fall by the wayside.

The ideal philosophy for safe operation of OCS platforms, of course, is that every operator will strive to "do the job right the first time," so that the role of the MMS inspector can be limited to verification. On the other hand, given the importance of safety devices in protecting life and property, what might be termed "micro-inspection" can lead to an operator attitude that the MMS inspector will detect anything that is wrong and that, in the absence of overt indication of a problem, the operator can concentrate on production.

Responsiveness to Other Evaluation Considerations

Maintenance of Public Confidence

An important element in the effectiveness of a federal regulatory program dealing with safety is the degree to which the public has confidence that its interests are being safeguarded. This is particularly critical with respect to OCS activities in light of the federal government's interest in realizing the substantial revenues to be derived from OCS development. On the Gulf Coast the question of public confidence in MMS' role in the safety of OCS operations is a dormant issue. The gradual migration of the center of industry activity from shore to marshes and lakes and to the open sea, its current remoteness from population centers, and its pervasive importance to the Gulf Coast economy all have combined to make the attendant risks an accepted part of the local scene. On the Pacific Coast, by contrast, there is a segment of the population which is unalterably opposed to any offshore development. This opposition tends to override the issue of public confidence in the MMS inspection program as such; the consuming issue is development in general, not the nuances of addressing day-to-day safety problems. In these circumstances it cannot be said that there is substantial public interest in the specifics of the MMS inspection program. In Alaska, on the other hand, the high level of public concern and threatened litigation resulted in lease stipulations responsive to public concerns which, in large part, have shaped the inspection program there.

Use of Resources

This study was prompted primarily by MMS's concern regarding the efficient utilization of government resources. The committee agrees that there are problems involving the utilization of the MMS inspection staff. A utilization[3] percentage on the order of 59 percent (Pacific) and 70 percent (Gulf of Mexico) of compensable time is unacceptable on its face. Some inefficiencies appear to result from MMS's inspection procedures. The substantial time spent in hurried records checks while at a facility would be better spent in direct inspection activity, with intensive records review conducted ashore, either before or after the facility inspection. Further efficiency in personnel utilization could be achieved by lengthening the work day to accommodate visits to more facilities. (Work rules would have to be changed to accomplish this, perhaps involving new legislation). The committee believes that changes of this nature could accomplish marginal improvements in productivity and decreases in transportation costs. It is questionable that they would result in any substantial impact on the safety of operations.

Assessment of the cost/benefit ratio of the present inspection program is not feasible on the basis of available data. That is, while MMS costs can be established with reasonable certainty, and while they can be applied to determine cost per inspection, cost per violation detected, etc., a measure of the safety impact of the program–in terms of such significant objectives as accidents, injuries, or pollution incidents prevented–is not feasible. It can be said that the cost per inspection is high;[4] that the cost per violation reported is even higher; that the number of major accidents is low; and that the fatality rate is similar to that of other heavy, labor-intensive industries (equivalent to that of the mining industry, for example). All in all, the true safety picture on the OCS and its relationship to the MMS inspection program cannot be defined quantitatively. (However, it is

---

[3]Time engaged in actual inspection and other assigned duties vs. total time the inspector is on duty; helicopter transit time is not included in the time credited to inspection.

[4]In FY 1987, average helicopter cost per inspection visit was $440 in the Pacific Region and $675 in the Gulf of Mexico (Rigg, 1988a, 1988b).

possible to make a qualitative assessment as to whether a particular change in the inspection program would result in a higher or lower cost/benefit ratio.)

Qualification and Training of Inspector Force

The committee noted that inspectors receive little formal education in safety practices and inspection techniques. (Their training places primary emphasis on testing of devices and much less on safety practices.) While on-the-job training (OJT) gives inspectors sufficient insight into the processes they are to inspect (and channels them into specializing in either production or drilling inspections), OJT can fail to give an adequate conceptual understanding of the goals of an inspection. Further, OJT does not give inspectors an understanding of a large variety of inspection techniques, particularly in making judgments about safe and workmanlike procedures. Despite the fact that to some degree inspectors are rotated, they and their individual habits and areas of emphasis during an inspection appear to become well-known to facility supervisors and their key personnel. This is particularly true in the Pacific region, where a relatively few inspectors deal with only a handful of platforms. The committee gained a clear impression that operators there attempt to anticipate the inspector's inspection pattern.

Identification of Safety Trends and Warnings

Identification of significant safety indicators and better data collation are needed to provide a clearer picture of safety problems on the facilities, so as to establish trends. While MMS maintains that the compliance record is a primary factor considered in selecting facilities for unannounced inspections, the committee saw indications that transportation costs and scheduling may be a more significant factor in the field. Moreover, a violation record involving no more than the failure of a device to trip at its specified set point or the failure of a check valve to seal within the small tolerances specified by MMS, for example, does not necessarily translate into a poor safety record.

Further, until more directly relevant data can be collected and collated effectively, it will not be possible to assess the impact of safety-related changes introduced into the offshore oil and gas industry. Until then, improvements in equipment and process can be introduced only on an instinctual basis–not necessarily a bad procedure, but not the way to ensure consistent results.

Safety Performance Accountability

The present program does lend itself to a degree of safety accountability for individual inspectors, operating personnel, supervisors, and operators by measuring, for example, the PINCs inspected and INC/PINC ratios. However, this focuses the inspection process further on compliance with specific regulations rather than on the evaluation of safety attitudes and procedures that underlie the overwhelming number of events. That is, the data bases currently being put together by the MMS provide data to hold individual inspectors, operators, etc., accountable for compliance activity. The present approach does not incorporate the kind of accountability for safety that leads inspectors, operating personnel, supervisors, and operators to address safety as a system rather than an item-by-item checkoff.

Precedents

Compared to the state regulatory regimes governing offshore oil and gas drilling and production activity in their waters the MMS program is more rigorous, both as to the scope of regulation and the onsite presence of inspectors. (It should be noted that wells and facilities in shallow water generally are not as complex as those located in deeper water.) The committee was not able to develop data that presented a valid insight into the safety record of operators in state waters; but no evidence was apparent to suggest that there had been significant numbers of major accidents.

Practices in other nations were surveyed with a view to developing possible alternatives for adoption in this country. They ranged from an essentially laissez faire attitude in several developing countries to a system seemingly more pervasive than our own in the North Sea fields administered by Norway. The United Kingdom's system, though comparable to our own in substance, differed significantly in that the day-to-day oversight was in the hands of third-party inspectors (typically classification societies certificated for this purpose by the government.)

The Norwegian system places particular emphasis on the responsibility of the operator for maintaining quality control and the safety of operations. The Norwegian Petroleum Directorate monitors the operator's safety control system rather than the operator's activities (MIT, 1986).

A risk analysis, that focuses on personnel safety, which addresses hazards, their consequences, and possible methods for mitigating them is often employed in development of a system of internal control (i.e., operator's quality control) of safety. Risk analysis is practiced in Norway, but it is oriented more toward identifying risks associated with large-scale catastrophic events than routine workplace accidents. Risk analyses are provided by the licensee in the concept safety evaluation conducted by the Norwegian Petroleum Directorate as part of the initial project approval. Subsequent platform modifications may invalidate this risk analysis as noted in a review of Norwegian and British uses of risk analysis in offshore operations (MIT, 1986). (See Appendix H for a description of the Norwegian and British systems.)

ALTERNATIVE 1: INCREASED INSPECTION ONSITE BY MMS

This alternative, mirroring as it does the present program, was analyzed solely with respect to the potential impact of "more of the same."

The negative elements of the present program probably would not be exacerbated, with the possible exception that an increased MMS presence on Gulf of Mexico platforms would lead operators to place even greater reliance on an expectation that MMS personnel will detect anything that is "wrong." In this event there would be a further clouding of the fact that the operator has ultimate responsibility for safety and a corresponding diminishment of operator safety consciousness.

While in some cases the increased presence of enforcement personnel has a positive influence on public confidence, it seems unlikely that this step would be significant with regard to public perception of the MMS's effectiveness in ensuring the safety of offshore operations, given the limited public awareness of the nature of the agency's activity. Nor would such an increase be an effective use of personnel resources as the program is presently structured. Additional personnel would be needed, as would additional funds for transportation services. Since the only benefit would be the witnessing of more of the same kinds of tests and drills, whose correlation to promoting safety the committee questions, the committee sees no merit in this alternative.

This assessment has been reached notwithstanding the fact that the gathering and evaluation of additional data has been urged by the committee as a necessary precursor to more effective inspection activity. However, the type of data that is needed is not an increased flow of reports of INCs, but a flow of information that defines more broadly the day-to-day safety deficiencies being encountered on offshore facilities.

ALTERNATIVE 2: INSPECTION OF A SAMPLING OF PINCs DURING ANNUAL INSPECTIONS, AND INCREASED SPOT INSPECTIONS

While rejecting the concept that an escalation of the MMS's present program would be beneficial, the committee nevertheless was struck by the fact that, at all levels, the agency and the operators believe that the MMS inspection presence promotes industry safety awareness. Some of this belief is based on the perceived quality and scope of MMS oversight; however, other factors contribute more emphatically to the consensus. Many operators who are intent on conducting safe operations through a conviction that it is economically sound practice express the feeling that the credibility of the industry's commitment to safety would be threatened by the actions of the less scrupulous operators if the possibility of an MMS spot inspection were to be reduced. Employees concerned about a safety hazard that management is not addressing to their satisfaction want to be able to aim an inspector in the "right" direction. MMS inspectors believe they bring a different viewpoint, which frequently enables them to detect problems that are overlooked by personnel preoccupied with the facility's ongoing operations.

This strong predilection for an MMS presence, which the committee shares, led the committee to explore ways in which to enhance the effectiveness of MMS inspections at present personnel and budgetary levels.

The committee examined whether the utilization of sampling techniques (see Appendix G) could enhance the effectiveness of the MMS inspection program. Sampling theory is a widely accepted technique for improving the efficiency of inspection programs. However, it is highly dependent on a properly developed sampling plan designed to verify the data recorded by the party being inspected. The committee believes that a valid sampling plan can be developed for witnessing tests of selected items on the PINC list during inspection of OCS oil and gas production facilities. However, better use needs to be made of the facility records, which represent considerable time and effort expended by the operators on conducting periodic tests to meet MMS requirements, as well as on maintaining the prescribed records. Review of the records in greater depth before an onsite inspection could focus the inspection more sharply. With proper controls, the records could take the place of many MMS-witnessed tests and provide the basis for developing the sampling protocols.

This alternative could lead to a program based fundamentally on the operator regularly performing the scheduled inspections, followed by selective inspections performed by MMS personnel utilizing a sampling plan. Making the tests performed by the operating personnel a formal part of the facility's MMS inspection program should foster a secondary benefit of extending safety involvement to all employees; it should help to focus attention on the fact that safety is derived from proper operating practices, rather than being the result of inspections. Integrating the efforts of the MMS inspectors with those of the operators should reinforce the precept that safety is the responsibility of the operator; if no more than a shift in attitude results, the outcome should be a level of safety higher than that achieved with the present system.

At the heart of this alternative is the commitment of time to an increased number of unannounced spot inspections, which will enable MMS inspectors to observe operating procedures more frequently and assess safety attitudes more accurately. While the committee is confident that the use of sampling techniques in conducting onsite tests and inspections can provide oversight of offshore operations that meets the objectives of the OCSLA, which is superior to that being achieved using present procedures, it also recognizes the possibility that questions could be raised as to whether sampling meets the specific OCSLA mandates relating to announced and unannounced inspections. The committee takes no position on the need for legislation in order to undertake a sampling program, but urges that legislation be sought if deemed necessary. The committee does note, however, that the Coast Guard has undertaken a sweeping change in respect to regulations it administers, based on an administrative interpretation of the OCSLA to the effect that Coast Guard personnel do not have to conduct all OCSLA inspections.[5]

As discussed in reference to other alternatives, the "public perception" issue does not appear to be an important one in addressing the acceptability of various inspection strategies. The alternative under consideration here does not involve a reduction in MMS activity offshore but a reallocation of effort, and accordingly should be relatively immune to public criticism.

The real advantage of the alternative lies in a more productive utilization of the current inspection force, both in terms of the activity it will undertake and in the better scheduling of its time and transportation services. Given a more efficient use of resources in terms of cost, the improvement in cost/benefits ratio still might be marginal unless there is a substantial improvement in the impact of inspections on the safety of operations on the OCS. This is the key element of this alternative. Improvement will follow only if the resources freed from having to witness tests are applied to evaluating events (accidents) and disruptions (mishaps) and to conducting safety analyses and if the inspection program is modified to assess safety-related attitudes. As was pointed out in Chapter 4, no change will result in more than marginally better safety without the identification and collection of more meaningful data. While a PINC sampling program can be initiated with minimal risk on the basis of presently available data, upgrading of data will have to be given high priority in order to refine the sampling program into one that will give the MMS, the industry, and the Congress confidence that the safety hazards of OCS operations are identified, understood, and addressed effectively. The following chapter elaborates on data needs and the reallocation of some inspector resources into data analysis to improve sampling and safety.

Although the committee believes that MMS inspectors are selected and trained to sufficiently rigorous standards to conduct their present inspection role satisfactorily, it has some reservations about the qualifications and training of inspectors with regard to the breadth of their understanding of inspection techniques, particularly of statistical sampling concepts. Many inspectors will require intensive training before they could be effective in a more sophisticated inspection regime, such as one built around safety analysis, sampling techniques, and audits of operator records.

Additional training also will be required to enable the current inspectors to assess operator attitudes toward safety and to identify procedures that involve poor safety practices. Learning to assess attitudes will be even more difficult than moving from witnessing all tests to witnessing a sampling of tests.

The focus of an inspection based on a sampling program is to determine the accuracy of the operator's records and to hold him accountable for his reported tests. In this sense, operator accountability is improved by this alternative program.

ALTERNATIVE 3: ANNUAL INSPECTION OF A SAMPLING OF FACILITIES AND INCREASED SPOT INSPECTIONS

The committee considered the application of sampling theory to identifying selected facilities for annual witnessing of tests while foregoing the annual announced inspection at the others. Depending on the level of confidence that could be established for this approach, a substantial

---

[5]The Coast Guard's regulations deal with the number and condition of lifesaving devices such as evacuation capsules, life rafts, and personal flotation devices, firefighting devices, both fixed and portable, and aids to navigation. In addition, the Coast Guard is responsible for general workplace safety. The Coast Guard's inspection program requires operators to perform (and report to the Coast Guard) all annual inspections. These inspection are augmented by a Coast Guard oversight program of random announced inspections, a headquarters analysis of inspection reports filed by operator and Coast Guard personnel, spot inspections of facilities with poor safety records or questionable reports, individual investigations of worker complaints, and investigations of casualties.

reduction in manpower and fiscal resources required to witness tests could be realized. An integral element of this alternative is a step-up of spot inspections to ensure that facilities not included in the announced annual inspection sample would receive at least one spot inspection during the year. The spot inspections would not have extended to witnessing a large number of tests.

This alternative might not result in literal compliance with the requirements of the OCSLA mandating an announced annual inspection and periodic spot inspections, therefore it might require legislative action prior to adoption. However, to the degree that an effective sampling plan can be developed, this alternative would meet the safety objectives of the OCSLA while freeing personnel, time, and funds for more productive activity.

Potentially, this alternative could have a highly positive impact on the safety awareness of operators and operating personnel who–to a greater degree than in the case of the PINC sampling program just described–could find themselves in the event of an accident having to explain a safety deficiency on a platform that had not been recently inspected by MMS. The reduction in the number of facilities receiving complete inspections by MMS inspectors conceivably could arouse public concern, which might require a public awareness program by MMS explaining the concepts underlying the selected-facility alternative.

As in the case of the PINC sampling alternative, key elements of the acceptability of the selected-facility alternative are a concurrent improvement in data collection and an aggressive safety analysis program. To the degree that these elements are accomplished, the forecasting of safety trends may well be enhanced, since it will be possible to look at safety data in terms of an integrated operating system (i.e., a facility). To be able to develop safety indicators that would identify a potentially problem-prone facility would add greatly to the efficiency of the MMS inspection process.

This alternative would place far greater demands on the analytical skills of the MMS inspectors, who would require extensive training in the use of modern risk assessment tools. It might be necessary to wait until a new generation of inspectors is in place before implementing this alternative. Future recruiting could accommodate this need over time so as to phase in personnel with these skills without displacing the present inspectors, who could continue to perform their present tasks.

Once in place, this alternative would afford MMS a high degree of flexibility in allocating its inspection resources to address problem areas. MMS responsibly could defer inspections of operators whose safety performance was superior, in order to concentrate on those encountering problems; or it could ease off inspections of categories of PINCs in which no problems were being encountered to concentrate on those that have posed safety hazards.

This alternative, placing as it does an additional burden of accountability on operators and operating personnel, might diminish the accountability of individual MMS inspectors and their supervisors for the performance of detailed inspection tasks. However, the overall accountability of MMS for the safety oversight of the industry should be enhanced by the fact that better data, more intensive safety analysis, and selective inspection activity should give MMS a better overall picture of the safety of the industry's operations (see Chapter 7).

In the committee's view this alternative represents an achievable long-term goal. At the moment, however, it is not feasible given the limited amount of safety data available, which would make formulation of a suitable selected-facility sampling program impossible. However, this alternative could flow naturally from the successful implementation of alternative 2, the PINC sampling alternative.

ALTERNATIVE 4: THIRD-PARTY INSPECTION WITH GOVERNMENT AUDIT

In this alternative, operators would contract with private organizations certificated by MMS to perform the annual and spot inspections now being performed by MMS inspectors. (At the operator's option the private organization could also perform the inspections now performed by the operator's own personnel.)[6] The third-party organization would maintain the records prescribed by MMS (which would be subject to spot audit by MMS) and submit copies of the inspection reports to MMS so that the quality of the inspections could be verified.

As in the case of some of the other alternatives described in this report, this alternative might require legislative authorization. Whether the objectives of the OSCLA would be better achieved by this or the present system is a matter of conjecture greatly influenced by the biases of the evaluator. Those supporting an increase in private-sector activity suggest that greater efficiency would result at no risk to attainment of the OCSLA's safety objectives; those who believe that government inspectors are the only ones sufficiently committed to the public interest to provide objective inspection services maintain that monetary savings would be doubtful and in any event not worth the ensuing degradation in the safety program. What is indisputable is that monitoring the performance of private inspection organizations in itself would pose significant difficulties. An MMS program to maintain a register of private-sector engineering consultants qualified to perform certifications of platform design, fabrication, and installation, which was undertaken some years ago, was abandoned due to difficulties in keeping it current.

It is hard to assess the impact that this alternative might have on the safety consciousness of operators. With adequate precautions to obviate conflict-of-interest situations, there is no reason to believe that third-party inspectors would not conscientiously carry out their duties so that the compliance element of the inspection process would be unchanged. This fact in itself, however, provides the operator with the same kind of shelter that he now has when he successfully "passes" an MMS inspection. Thus, the committee believes that there would be negligible impact on safety consciousness; the tendency toward "compliance mentality" would not be corrected by this alternative. The drastic reduction in MMS presence on OCS facilities could raise public concern. Moreover, as noted earlier, there is strong sentiment in the industry, on the part of offshore operators and employees, as well as MMS employees, that the regular presence of MMS personnel has positive benefits on safety which should not be foregone.

With the reduction of their inspection duties, MMS inspectors would have time to perform the audit and safety analysis functions the committee described earlier as providing the MMS with better insight on operational safety on the OCS. In this case, however, they also would be required to conduct audits of the private inspection organizations. These functions in the aggregate could be performed with greater scheduling flexibility than direct inspections and would allow MMS to make more effective use of its personnel resources (although there might be some degradation in the inspectors familiarity with the practical aspects of OCS operations).

The overall cost of this alternative probably is higher than that of the other alternatives described in this report, since it introduces a new group of inspectors into the process (paid directly by the operator, or as an add-on to the MMS budget, depending on the variant adopted). Some reduction in the number of MMS inspectors would be possible in the case of either variant, but some redundancy would remain and the MMS audit of inspectors would likely involve higher cost per inspection. It is questionable whether there would be a sufficient improvement in effectiveness to offset the additional costs involved.

Paralleling, as it does, the substantive aspects of the present program, the alternative would produce no direct improvement in the development of data useful in improving safety trend analysis. However, adoption of this alternative would not foreclose a transition into one of the other alternatives described (e.g., alternative 2), in which such benefits are realized. In fact, such a transition might be facilitated since the MMS personnel would be able to concentrate their attention on program development without the distraction of day-to-day involvement in facility inspection.

---

[6]As discussed in Chapter 4, a variant would be that MMS would hire the third-party inspection services directly. The committee believes that the following analysis is equally pertinent to this variant.

Adoption of this alternative would involve difficult transition problems in training the MMS inspectors to perform audit functions relating to the activities of third-party inspection organizations. These skills are less related to the background of the inspectors than those involved in witnessing tests, and might require protracted retraining.

The use of private-sector inspectors probably would provide some additional flexibility to MMS in modifying the inspection program in the face of changing circumstances. The degree to which this flexibility would be meaningful depends entirely on the ability of MMS to develop better data collection and evaluation processes.

A fundamental challenge in adopting this alternative is development of the audit plan for overseeing the activity of the third-party inspection organizations. Without an effective scheme allowing MMS to follow up on the performance of the private-sector inspectors, some operators could experience a very rapid deterioration in the safety of their operations. With government inspectors no longer involved in direct inspection activity, it might be difficult for MMS to ensure that its mission was being carried out effectively.

Third-party inspection programs are not commonplace in the federal regulatory scheme, although they do exist. For example, there is statutory authority for the Coast Guard to delegate a wide array of marine safety functions to the American Bureau of Shipping (a classification society of the type utilized by the United Kingdom to conduct its inspection program for offshore oil and gas operations). The Coast Guard has made numerous delegations under this authority, but it has not delegated operational inspections. None of the states' programs involve third-party inspectors.

A transition to a third-party inspection program with government audit would not of itself result in an improvement in the safety of offshore operations, nor would it result in a more efficient inspection scheme. It presents difficulties for the MMS in conducting oversight that it may not be able to handle without substantial near-term disruption, during which safety oversight might be seriously compromised.

ALTERNATIVE 5: SELF INSPECTION

The thrust of the views presented to the committee was that hazards presented by offshore gas and oil operations are too significant to consider an inspection program that does not involve direct government oversight.

Although, as in the case of the other alternatives, this alternative involves extensive assessment of operator data in meeting the safety objectives of the OCSLA, there is a basic difference: the others still rely on a regular (if more limited, in some cases) oversight by MMS inspectors or their proxies to verify the operator's inspection activity and his records. Full self-inspection would involve a loss of the motivational benefits of a governmental presence referred to in Chapter 4. It certainly would risk a loss of public confidence in the safety of OCS operations, and might be seen as an abandonment of governmental responsibility.

Substantially more MMS resources would be available for spot inspections and safety analysis activities, as noted in Chapter 4. But it is likely that the collection of valid, detailed data needed for those analyses would become more problematical, as the frequent direct contact between MMS inspectors and operating facilities and personnel was lost. For the same reason, the day-to-day performance of OCS operators would become more difficult to ascertain. In the absence of government inspection, the operator's accountability for poor performance seemingly would be self-evident. Nevertheless, after a major accident, MMS's discharge of its regulatory responsibility could be questioned by the public.

RESULTS OF THE ASSESSMENT

Having assessed the present MMS inspection program and five alternative programs, the committee made a determination that alternative 2–inspection of a sampling of PINCs during annual inspections, and increased spot inspections–offers the greatest relative benefits. It provides an opportunity to improve the efficiency of the inspection program while retaining a substantial government presence in this area of vital public interest. It reduces the tendency toward a "compliance mentality" on the part of operator personnel while focusing the inspection on the items of hardware and procedures that are most significant from a safety standpoint. It opens the way for the program to evaluate the aggregate safety of operations on all platforms.

This alternative inspection program would result in fewer total inspection man-hours required for annual inspections, as well as better targeted spot inspections. The committee recognizes that this increased efficiency could be regarded as a basis for manpower reductions. However, use of the freed resources for additional spot inspections and a program of systematic safety analyses that are needed to address recurrent safety problems is an integral element of this alternative. The necessary resources for an ongoing safety analysis program and an enhanced spot inspection program would be made available by having MMS inspectors witness only a limited number of tests during the annual inspections of production facilities, in lieu of attempting to witness 100 percent of the tests in the PINC list as is now done. The alternative also would result in lower transportation costs since the number of days spent at each facility would be reduced.

While some uncertainties would be involved because of shortcomings in the MMS data base, the transition to the new inspection program could be undertaken at an acceptable level of risk concurrently with the development of analytical methods to improve the data base to make it more useful.

Alternative 3–annual inspection of a sampling of facilities, and increased spot inspections–represents a natural and perhaps desirable extension of alternative 2. The committee considers this step to be too ambitious a change to undertake at the present time, given present limitations in the data and expertise needed to confidently select facilities for inspection. But as those limitations are overcome through the implementation of alternative 2, the selected-facility alternative would provide even more opportunity for targeting inspection resources efficiently, effectively, and intelligently at the points on the OCS where attention is needed most. Thus, implementation of this alternative represents a desirable long-term goal, recognizing that legislation may be required before it can be undertaken.

7

REDIRECTION OF INSPECTOR ACTIVITY TO ENHANCE THE INSPECTION PROGRAM

An integral element of the modified OCS inspection program recommended by the committee is enhanced collection and analysis of safety-related data. The data collection effort must go considerably beyond the present Minerals Management Service (MMS) Offshore Information System in a three-fold purpose:

1. to support the development and implementation of sampling plans for potential incidents of non-compliance (PINCs) (and later perhaps, OCS facilities),
2. to permit systematic targeting of spot inspections, and
3. to support a variety of continuing safety analyses to be used to improve safety and environmental protection on the OCS.

All three activities are essential to an ongoing "risk assessment and management" program. Data and analytic tools used for sampling plans and spot inspection purposes will center around the identification of safety trends, while those used for the ongoing safety analyses will be more varied.

Carrying out this safety analysis program should not require additional resources; it can be accomplished by reprogramming of existing inspector resources that will be released through the increased inspection efficiency inherent in the recommended alternative inspection program. Given the available work force, how should the inspector activity be redirected? The answer requires an evaluation of the tasks that would effectively serve safety and environmental protection needs on the OCS. The tasks described below could and should be carried out as an integral part of any OCS inspection program. Indeed, the committee believes that one major deficiency of the present inspection program is that it does not incorporate these functions.

TASK REDIRECTION

For the short term, the redirection of tasks should focus on the following objectives:

- determining the acceptability of safety performance of all operators,
- improving the verification of the effectiveness of regulatory requirements, and
- improving the process by which emerging risks are identified.

Determining the Acceptability of Operator Safety Performance

Two aspects of this objective must be addressed: (1) currently less-than-acceptable operations,[1] and (2) currently acceptable but deteriorating operations.[2] First, the focusing of inspection resources on operators who are not conforming to regulations or lease requirements–either in general or in particular aspects of their operations–and who have less than adequate safety and environmental risk control programs is and should continue to be a priority concern of the MMS. These marginal or unsatisfactory operations should be subjected to MMS surveillance on a priority basis, since deficiencies in risk controls clearly increase the probability and potential consequences of accidents leading to fatalities, personal injuries, property damage, and/or environmental harm. (These are but a few of the more obvious consequences that affect public confidence regarding OCS operations.) Next, deteriorating operations need to be identified and corrected before they deteriorate to the point that they create unacceptable safety or environmental risks.

Identifying Less-than-Acceptable Operators

The current process by which less-than-acceptable operators are identified is unclear. The committee was given no documentation or inspection results that identified less-than-acceptable operators. The design and implementation of a system using modern statistical data analytic sampling techniques to uniformly identify these operators is essential. An appropriate system would incorporate a variety of records, in addition to inspectors' opinions. This would require periodic records monitoring and review of a type that would shift a greater portion of the inspector's work hours to records inspections in the operators office. Real cost saving could result, because travel costs to management locations ashore, where records are maintained, would be lower and could be scheduled more efficiently than travel to and from the offshore facilities. There would be the further advantage that the inspectors could observe the development and implementation of the operators' safety and environmental risk control systems at the management level, from a perspective that is not now observed directly by inspectors.

Identifying Deteriorating Operations

This aspect is more difficult to address. There was no evidence that MMS regions and field offices presently are using operator inspection histories systematically to address this problem. Given the present inspection schedules for inspectors, especially in the Gulf of Mexico area, only occasionally will inspectors note deteriorating performance or get general impressions that indicate it is occurring. In the absence of formal guidelines, the MMS inspection system at best can rely only on the unstructured judgments of individual inspection personnel, with the inconsistency and subjectivity that individual judgments entail.

---

[1]An operator may have acceptable operations in one area and poor operations in another area due to varying operational environments or different personnel, for example.

[2]"Deteriorating operations" refers to the deterioration (notably piping systems) of older platforms due to age or due to the fact that depletion of the wells has made standard maintenance practices uneconomical. It also refers to deterioration of operating procedures and management attitudes affecting safety.

A demonstrably more effective approach could be achieved through analysis. The records monitoring system can and should be designed to detect degradation of safety and environmental risk control systems.

Data Collection and Review

The inspection data system design should not be limited to scheduled inspection of operator records at operators' operations centers. Spot requests by MMS for operational, inspection, or test data, by telephone or radio communications with manned facilities should be provided for. These unannounced data requests as well as interviews of key individuals who work on the facilities would provide important information on the actual state of safety systems. The selection of key persons to be interviewed has to be carefully considered to enhance the probability of receiving reliable information. At present, there is no established mechanism by which MMS can obtain data or talk to individuals other than during the highly structured onsite inspections.

Where data from an individual operator suggest there has been a deterioration of safety systems, the inspection activity could be expanded to include onshore interviews with responsible operating employees. While such information would be used primarily to identify specific deficiencies in existing safety and environmental risk control systems (and thus provide an additional data source), it also would provide a basis for shaping future inspection tactics regarding that operator. These and other indications of deteriorating safety operations would then lead to unannounced inspections of the kind now performed by MMS inspectors; but those inspections would be guided by the new safety analysis program toward specific ameliorating actions.

Verifying the Effectiveness of Regulatory Requirements

Efforts need to be devoted to verifying the effectiveness, (i.e., appropriateness and adequacy) of the MMS permit approval process for drilling, production, workover, and well shut-in activities, as well as the related regulations and training requirements. More feedback from actual operational conditions and experience is needed. Conceptually, the permit approval process is based in part on postulated failure modes and the installation of equipment and procedures to deal with them. However, the committee observed no evidence that this information was documented in a form that could be used by MMS inspectors to perform system evaluations during platform inspections, rather it was useful only for go/no-go compliance observations of hardware. Some kind of real-world feedback is needed to verify that regulations and training requirements are effective in controlling safety and environmental risks.

Opportunities for Acquiring Verification Data

Inspections should be used to provide MMS with feedback data in all the foregoing areas. The present PINC list encourages inspectors to look for specific items of noncompliance rather than accident-producing conditions or conditions promoting human error. The committee could find no clear relationship (and certainly not one that was documented) between accident prevention–on which approval, regulatory, and training requirements should be based–and the check items on the PINC list.

Further, if inspectors were provided with documentation of the basis for the approvals, as well as for regulatory and training requirements, they could make more informed observations of safety-related activities and procedures on platforms. The information would allow them to identify conditions, activities, and attitudes that might degrade safety and determine how well they were being controlled by the operator's safety programs and personnel. The observations could cover procedures and their execution as well as hardware and its performance.

Information Flow and Access Considerations

Information flow channels were not examined extensively, but it appeared to the committee that utilization of information from inspections was not effective due to the lack of formal criteria for its incorporation into the regional data base.

Criteria and procedures for documentation and distribution of inspectors' data either verifying or raising questions about approvals, as well as regulatory or training requirements, are needed within MMS as a part of a redirected inspection effort. Also, provisions need to be made to encourage inspectors to report candidly their subjective observations and interpretations.

Identification of Emerging Safety Risks

Redirection of Inspection Activity

Use of the inspection staff to detect emerging safety risks is another area requiring greater emphasis by MMS. As noted earlier (see Chapter 3), the current compliance inspection program is focused on detection of single-point hardware failures addressed by specific regulations. The practical effect of this "compliance" emphasis is that inspectors limit their inspection to the PINC list, the detection of non-compliance with items on it, and the enforcement actions specified in it.

This approach provides little or no room or incentive for inspectors to uncover and define risks that may not be addressed in the approvals, regulations, training, or PINC list requirements. It is not clear how MMS identifies emerging risks. In any case, this function is not included in the current position description for inspectors (see Appendix E). The inspectors, who are in daily contact with current operations on the OCS, should be "lookouts" for emerging risks, and would have the time to so act if their activity devoted to PINC list compliance were reduced; they will need retraining to be effective in this role, however. The detection of emerging risks and potential accident or mishap scenarios should be made a specifically defined function of the inspection effort.

Broadening of Reporting and Investigation Criteria

Additionally, the scope of the MMS investigations of certain events (accidents) should be extended. First, the definition of a mishap in MMS regulations should be changed to include: "near misses," any drilling or production disruption, and events which prompt an MMS inspector to shut down operations, such as the failure of a mud level gauge in a drilling operation.

Second, operators should be required to record and investigate all mishaps, and to keep a log of the mishaps and the investigations, which would be made available to inspectors on request. This procedure would permit MMS to learn about risks on a broader basis than just major events (accidents) and mishaps that happen to occur on the platform when an inspector is present. Inspectors would thus be in a position to review with the operator while onsite any prior occurrence in which safety-critical or single-point-of-failure components or procedures broke down or failed, and also to identify the likely consequences of a similar event or mishap and how to correct the problem, i.e., any changes made by the operator as a result of the investigation.

Dissemination of Investigation Results

Information obtained in an event or mishap investigation should be disseminated to MMS units with a need to act, and to operators, in a manner similar to the safety alerts now used for a narrower purpose. The event and mishap data should be more descriptive, relating the event or mishap to risks, and the reports should be issued frequently. Since additional information would benefit both MMS and the operators, the change should be well received by the operators. From the standpoint of MMS, this flow of information should also be beneficial in helping to eliminate "compliance mentality" and in developing a cooperative "safety attitude" among OCS operators.

Action on Reports of Emerging Safety Risks

Action on the information developed by these redirections of the inspection staff activity might be reflected in changes in approvals, regulations, or training requirements. To ensure that the information is utilized, a process should be established to provide inspectors with feedback about the action taken on their reports.

INSPECTION FORCE CAPABILITIES

Observation of MMS inspectors in the field indicates that redirection of their efforts described above would require some retraining. To accomplish records review using statistical methods, for example, inspectors would require training regarding effective procedures and analytical methods to achieve the desired results. To gather information for validating MMS requirements, retraining would be necessary to incubate an understanding of the concepts underlying the hypothetical or real accident scenarios on which regulatory requirements are based. Most of the MMS staff have had accident investigation training that is sufficient to deal with an expanded investigative function. The committee believes that with a manageable training effort the current inspectors would be for the most part responsive to and capable of adapting to the new tasks. As noted in Chapter 6, it may be necessary to recruit some inspectors over time to address the analytical needs of the enhanced program.

PERFORMANCE MEASUREMENT SYSTEMS

The committee was not able to obtain a clear picture of the measures used to assess performance of inspection program managers, supervisors, and inspectors. Apparently, the number of inspections completed is the primary criterion for assessing managers' and supervisors' performance. For individual inspectors, it appears that performance assessment methods and criteria may differ from region to region and district to district. The criteria for inspection force performance measurement criteria should be better delineated.

Expansion of the scope of inspection activity, as described above would provide measurable benchmarks for assessing inspectors' and managers' performance. The identification of less-than-adequate operator compliance using data analysis also would enable the MMS to measure the effectiveness of the districts and regions in detecting unacceptable operations and fashioning corrective programs for operators to undertake. While measuring the number of violations cited is not important for its own sake, what is measured gets attention. Thus it is important to place a positive focus on improvements brought by inspector intervention. Accordingly, the introduction of means of measuring the work produced under the redirected efforts should be implemented by MMS before the changes are placed in effect.

8

FINDINGS, CONCLUSIONS, AND RECOMMENDATIONS

In response to the request of the Minerals Management Service (MMS), the committee evaluated the present outer continental shelf (OCS) inspection program and identified ways to improve it. The committee's assessment of the current inspection program and five alternative programs (alternatives 1 through 5) against a variety of considerations led to the following findings, conclusions, and recommendations.

FINDINGS AND CONCLUSIONS

1. The current inspection program satisfies the literal requirements of the Outer Continental Shelf Lands Act (OCSLA) for annual and unannounced inspections of each OCS facility. In this sense, it fulfills MMS's statutory mission. On the other hand, the OCSLA carries a broad mandate to promote safety of life and property and to protect the environment. The committee believes that by this measure the MMS effort is not totally successful, particularly in the light of greater public expectations regarding safety and environmental protection that have emerged since the OCSLA was amended in 1978. In any event, changes in the OCS operating environment–including aging platforms, more complex systems and operations, activities in deeper water at greater distances from shore, and changing characteristics of operating companies–have created a need to upgrade the MMS inspection program in order to sustain in the future the generally good safety and environmental record of the past.

2. The core of the present MMS inspection program is ensuring compliance with a large number of hardware-oriented requirements compiled in a list of "potential incidents of non-compliance," the PINC list. The MMS's overriding emphasis on this list tends to lead operators to concentrate their safety activity on items they expect the MMS to inspect, i.e., the items on the PINC list. The net result is that to some extent the program fosters an attitude that "compliance equals safety." This "compliance mentality" is of concern to the committee because it can diminish the operator's recognition of his primary responsibility for safety.

3. There is a considerable degree of variability among offshore operators regarding the amount of emphasis that each places on safety programs. The safety attitude of some operators is reflected in a failure to institute adequate safety programs, to maintain important safety equipment systematically, and to encourage good safety attitudes and practices among operating personnel.

4. The current MMS inspection program does not incorporate satisfactory means to specifically identify, measure, or react administratively to the evidence of insufficient emphasis on

safety programs, where it exists. Nor does it lead MMS to vary the level of its inspection activity to focus greater attention on facilities operated by lax operators. Presently, PINC G-400, dealing with whether operations are conducted in a "safe and workmanlike manner," provides the only vehicle for MMS inspectors to cite subjective observations regarding poor operator attitudes toward safety. In itself it is an insufficient basis for remedial action.

5. The majority of accident events occurring on the OCS in a representative year (1982) were related to operational and maintenance procedures or human error that are not addressed directly by the hardware-oriented PINC list. In its close examination of the 1982 data only one of the 11 events leading to fatalities were found by the committee to be related even remotely to a hardware PINC. None of the fatalities was caused by the failure of a safety device whose testing must be witnessed by MMS inspectors–although witnessing tests currently is the principal inspection activity of the MMS inspection program. Similarly, of 263 events in 1982 involving injuries, fires, and pollution on or from production facilities, only two probably were due to failure of a safety device.

6. Most accident events occurring on the OCS result only in minor personal injury, small flash fires, and small oil spills. MMS focuses its attention primarily on "major accidents"–those that result in significant injuries, deaths, and large spills. However, low-level mishaps often are precursors to major disasters. To reduce the risk of major disasters it is necessary also to study and understand the root causes of low-level mishaps.

7. As a previous Marine Board report pointed out (National Research Council, 1984), the MMS needs to improve its data collection and safety analysis effort to meet modern standards for a safety program. While MMS has made progress in data collection over the past several years, its organization and analysis of those data is not sufficient to meet the needs of a safety program which goes beyond inspection of specific devices and controls. The present program incorporates no mechanism or analytical basis for systematically upgrading safety requirements for OCS operations. For example, the committee found no indications that the MMS

- analyzes data to identify safety trends,
- collects data consistently across operators and facilities that would permit such analyses,
- documents operator safety histories, or
- cross-references PINCs and incidents of non-compliance (INCs) to events (accidents).

8. The present checklist approach to inspection and testing does not encourage MMS inspectors to be alert to, or to report on, emerging safety risks due to changing technologies, procedures, or operating environments. Yet these unrecognized risks could present significant threats to safety on the OCS, and an effective inspection program must seek to identify them, using guidelines which will preclude arbitrary inspection activity.

9. Based on these findings, the committee concludes that more inspections of the type that currently are being conducted (i.e., alternative 1) would yield no substantive improvement in safety. Such an expansion would require additional personnel and involve greater travel costs, while increasing the tendency of operators to abdicate their safety responsibility to the MMS inspectors.

10. On the other hand, the presence of government inspectors on the OCS is important for conveying a sense of oversight and for providing impetus to marginal and inexperienced operators to meet federal safety standards. Third-party inspection by private sector contractors (alternative 4) would not diminish and would probably increase the tendency of operators to abdicate safety responsibility to the inspecting organization. It would greatly alter the role and function of MMS inspectors, requiring extensive retraining and new criteria for hiring MMS personnel. It would present a strong potential for conflict of interest on the part of the third-party organization. And it

would almost certainly increase the cost of the inspection program. Self-inspection (alternative 5), while it would pinpoint the operator's responsibility, would be unsuitable because the MMS oversight function would be too tenuous. In the event of a major accident MMS would be vulnerable to charges that it had abdicated its responsibility.

11. The present inspection program can be modified to enhance its effectiveness at present levels of personnel and funding. The committee concludes that the effectiveness and efficiency of inspections to ensure compliance with regulations could be substantially improved by immediate adoption of an inspection program (alternative 2) utilizing modern sampling techniques to select items for inspection and, potentially, to select facilities to be spot inspected on a priority basis. This approach would free inspector resources for necessary complementary activities not now being carried out that would improve, to a significant degree, the content and implementation of the inspection program and thus meet the broad mandate of the OCSLA.

RECOMMENDATIONS

1. The committee's overall recommendation is that the MMS should maintain the presence of an inspector force on the OCS. The resource level of the inspection program should not be reduced. However, those resources should be utilized more effectively in an inspection program that permits more controlled and deliberate assessment and management of safety risks on the OCS.

2. To that end, the committee recommends adoption of a modified inspection program (alternative 2) for production platforms in which a facility-specific sampling of PINCs is selected for inspection during the annual inspection, based on a formal sampling plan. Sampling plans would be developed for each facility on the basis of a prior review ashore of records submitted by the operators as well as the results of prior inspections and industry-wide experience. Some of the inspector resources made available by witnessing fewer tests should be redirected toward increased spot inspections, instituted systematically on the basis on analysis of inspection results, operator safety histories, interviews with key operator personnel, and analysis of data to identify emerging safety problems and general safety trends.

This alternative program would require the operators to perform and record in a prescribed format all the scheduled inspections themselves (both those specified by MMS and those in the operators' own facility inspection program) with only selective verification inspections being performed by MMS in accordance with a sampling plan. Ultimately, as experience and confidence in the validity of the sampling plans is developed, the inspection program might be modified further to use sampling theory to select OCS facilities for a full annual inspection, as described in Chapter 4 (alternative 3). (Implementation of the latter alternative might require legislative action to amend the OCSLA requirements regarding annual inspections.)

3. The foregoing recommendation applies only to production operations. The committee recommends that the MMS's current "high visibility" program of frequent and comprehensive inspections of facilities engaged in drilling and workover operations, and the drilling and workover operations themselves, be continued because of (1) the high frequency of events per unit for these facilities as compared to production facilities, and (2) the large population of workers on each facility engaged in drilling and workover operations. Moreover, MMS should introduce the upgraded data collection and safety analysis features recommended by the committee for production platforms into its inspection program for drilling operators.

4. The committee recommends that the MMS begin collecting the types of data that will provide the basis for facility-specific sampling plans. This includes data on operator safety histories (consistently across facilities and operators), on PINCs and INCs as a function of accident events

and mishaps, on operator inspection and test results over time, on training programs, and on the condition of safety and environmental control systems. The safety analysis program should include monitoring and review of operator records ashore, and analysis of data and subjective observations to uncover safety trends.

5. The committee recommends that MMS place its primary emphasis on detection of potential accident-producing situations–particularly those involving human factors, operational procedures, and modifications of equipment and facilities–rather than scattered instances of non-compliance with hardware specifications. One reason for improving the inspection process should be to provide data that can be used to verify the relationship between items on the PINC list and accident prevention.

6. The committee recommends that MMS encourage its inspectors to uncover emerging kinds of safety risks and changing risks on OCS facilities. The position description, job assignments, and reward structure for MMS inspectors should be modified to reflect the importance of uncovering and reporting safety risks. An important step is to extend the definition of a "mishap" to include near misses, i.e., drilling or production disruptions, and events that prompt the operator or an MMS inspector to shut down operations and require investigation of these less serious occurrences as well as events (accidents). Information on events and mishaps should be disseminated in a manner similar to safety alerts, and should be used, where appropriate, in formulating changes in approvals, training requirements, and/or regulations.

REFERENCES

American Petroleum Institute. 1974. Analysis, Design, Installation and Testing of Basic Surface Safety Systems for Offshore Production Platforms. RP 14C. Washington, D.C.: API.

Arnold, K. E. and M. Stewart. 1989. Surface Production Operations. Volume 2: Design of Gas Handling Systems and Facilities. Houston, Tex.: Gulf Publishing. 1989.

Baker, R., 1979. *A Primer of Oil-Well Drilling.* 4th edition. Austin, Tex.: Petroleum Extension Service, University of Texas at Austin.

Bureau of Labor Statistics. 1986. BLS Bulletin 2259. Washington, D.C.

Massachusetts Institute of Technology, Center for Policy Alternatives. 1986. Risk Analysis in Offshore Safety and Environmental Management. Floyd R. Tuler, Christopher T. Hill, David W. Cheney, George R. Heaton, Jr., Sandia National Laboratories Contract No. 50-1610, project supported by the Minerals Management Service, Department of the Interior, Washington, D.C.

National Research Council, Marine Board. 1979. Implementing Best Available and Safest Technologies for Oil and Gas. Washington, D.C.: National Academy Press.

National Research Council, Marine Board, Committee on Safety and Offshore Oil. 1981. Safety and Offshore Oil. Washington, D.C.: National Academy Press.

National Research Council, Marine Board, Committee on Outer Continental Shelf Safety Information and Analysis. 1984. Safety Information and Management on the Outer Continental Shelf. Washington, D.C.: National Academy Press.

National Research Council, Marine Board, Committee on Disposition of Offshore Platforms. 1985. Disposal of Offshore Platforms. Washington, D.C.: National Academy Press.

Rigg, J. B. 1988a. Memorandum from John B. Rigg, Associate Director for Offshore Minerals Management, MMS, to the Regional Harbor, Pacific OCS Region, 23 February 1988.

Rigg, J. B. 1988b. Memorandum from John B. Rigg, Associate Director for Offshore Minerals Management, MMS, to the Regional Director, Gulf of Mexico OCS Region. 23 February 1988.

Rigg, J. B. 1988c. Memorandum from John B. Rigg, Associate Director for Offshore Minerals Management, MMS, to the Asst. Director for Program Review, MMS, 29 April 1988.

U.S. Department of the Interior, Minerals Management Service. 1984. Minerals Management Service Handbook. Washington, D.C.: U.S. Government Printing Office. July 17.

U.S. Department of the Interior, Minerals Management Service. 1986a. Minerals Management Service Manual. Washington, D.C.: U.S. Government Printing Office. September 3.

U.S. Department of the Interior, Minerals Management Service. 1986b. Managing Oil and Gas Operations on the Outer Continental Shelf (MS-646). Washington, D.C.: U.S. Department of the Interior.

U.S. Department of the Interior, Minerals Management Service. 1987a. Trip handout prepared for the Marine Board by MMS, Pacific OCS Region, 4 December 1987.

U.S. Department of the Interior, Minerals Management Service. 1987b. Pacific Summary/Index: June 1, 1986-July 31, 1987 (MMS 87-0078). Vienna, Virginia: MMS. 1987.

U.S. Department of the Interior. 1987c. MMS Budget Justifications, FY 1987, p. MMS-58. Washington, D.C.: U.S. Government Printing Office.

U.S. Department of the Interior, Minerals Management Service. 1988a. Accidents Associated With Oil & Gas Operations (OCS Report MMS 88-0011). Vienna, Virginia: MMS.

U.S. Department of the Interior, Minerals Management Service. 1988b. Federal Offshore Statistics: 1986 (OCS Report MMS 88-0010). Vienna, Virginia: MMS.

U.S. Department of the Interior, Minerals Management Service. 1988c. Gulf of Mexico Newsletter, 5 September 1988.

U.S. Department of the Interior, Minerals Management Service. 1988d. MMS: In Perspective 1982-1988. November. Washington, D.C.: U.S. Department of the Interior.

APPENDIX A

COMMENTARY ON FEDERAL MANAGEMENT OF OCS OPERATIONS

Active federal management of the development of oil and gas resources on the outer continental shelf (OCS) began in 1969. This management extends to imposing and enforcing measures to minimize the risks of development and to ensure the safety of offshore workers, the marine environment, and offshore installations. The current program of federal safety regulation and inspection derives from the Outer Continental Shelf Lands Act (OCSLA), passed in 1953 and amended in 1978.

Beyond the OCSLA, a panoply of federal environmental and safety laws affect OCS operations: the various Clean Air Acts; the Federal Water Pollution Control Act, as amended; the Rivers and Harbors Appropriation Act of 1899, as amended; the Ports and Waterways Safety Act of 1972; the Occupational Safety and Health Act of 1972; and the Natural Gas Pipeline Act of 1968 are among the more significant. The foregoing legislation is administered by a number of federal agencies, raising problems of coordination of activity; however, the MMS and the U.S. Coast Guard have primary jurisdiction over OCS activity under the OCSLA.

While both Minerals Management Service (MMS) and Coast Guard jurisdiction attaches to all vessels and structures used in drilling for or producing oil and gas on the OCS, the division of authority is fairly clear-cut: the Coast Guard regulates vessels and vessel-like platforms (principally mobile offshore drilling units [MODUs]) comprehensively and regulates fixed platforms primarily with respect to lifesaving and firefighting equipment; MMS regulates fixed platforms comprehensively and regulates all drilling and production activity on both fixed and mobile platforms.

In both cases the regulatory process is conventional: the basic regulatory authority in the enabling statute(s) is amplified by regulations and interpretive materials; compliance is exacted by the threat of civil penalties or the threat of immediate shutdown of operations (criminal sanctions are not a day-to-day factor in the regulatory process); and enforcement is accomplished through inspections conducted by government employees.

The underlying purpose of the OCSLA is to provide for the orderly and expeditious development of OCS resources. The original (1953) act recognized the need to conduct oil and gas operations safely. It emphasized that procedures and technology should be used which would minimize the likelihood of blowouts, fires, spills, and interference with other uses of the offshore waters. The 1978 amendments to the OCSLA, though prompted in large part by the national policy initiative for energy self-sufficiency, came at a time of intense public interest in environmental protection. As a consequence, environmental and safety considerations were addressed in very specific terms which gave them added policy emphasis. Thus, energy resource development is to be balanced by protection of the human, marine, and coastal environments; also, the development of new and improved technology to eliminate or minimize risk of damage to the human, marine, and coastal environments is encouraged. It was these amendments that also promulgated the statutory requirements for a minimum of one scheduled comprehensive inspection annually of each OCS facility subject to environmental or safety regulation, and additional periodic onsite inspections without notice.

APPENDIX B

U.S. COAST GUARD INSPECTION PROGRAM FOR FIXED OCS OIL AND GAS FACILITIES[1]

GENERAL BACKGROUND

In 1953, the U.S. Coast Guard was given the authority to issue regulations to promote the safety of life on artificial islands and fixed structures on the outer continental shelf (OCS) of the United States. The ensuing regulations contained requirements for inspections, construction standards, lifesaving and fire fighting equipment, operating procedures, casualty reporting, safety zones, and penalties. P.L. 95-372, the OCS Lands Act Amendments of 1978 (OCSLA 1978), assigned new authority and responsibilities to the Coast Guard. As a result, the regulations (33CFR 140-147 [Subchapter N]) were extensively revised to address the new responsibilities. (The revised regulations were published in the Federal Register by the Coast Guard on March 4, 1982.)[2]

OCSLA 1978 states that "The Secretary [of the Interior] and the Secretary of the Department in which the Coast Guard is operating shall individually, or jointly if they so agree, promulgate regulations to provide for

1. scheduled onsite inspection, at least once a year, of each facility on the Outer Continental Shelf which is subject to any environmental or safety regulation promulgated pursuant to this Act, which inspection shall include all safety equipment designed to prevent or ameliorate blowouts, fires, spillages, or other major accidents; and
2. periodic onsite inspection without advance notice to the operator of such facility to assure compliance with such environmental or safety regulations."

In the vast majority of instances the Coast Guard has had no reason to visit a fixed facility other than to conduct an annual inspection. Due to the magnitude of this task and the need to utilize available resources to respond to more critical safety areas, such as marine casualties and worker safety complaints, the Coast Guard has been unable to conduct annual inspections of all fixed facilities on the OCS. Recently, the Coast Guard further modified its regulations to require

[1]Prepared by the U.S. Coast Guard at the request of the committee.

[2]The U.S. Coast Guard inspection program for mobile offshore drilling units (MODUs) is more comprehensive. It is based on the classification of these units as vessels and addresses all on board systems in addition to industrial drilling and production equipment.

that annual inspections of fixed facilities be conducted by the facility owner's or operator's personnel or by a third party employed by the owner or operator.[3]

Under this program, the owner or operator certifies to the Coast Guard that the inspection was performed and states the results of the inspection, listing any deficient items. The efforts of Coast Guard marine inspectors then are focused on conducting unannounced oversight inspections of randomly selected fixed OCS facilities with resident personnel. Additionally, those fixed OCS facilities that are the subject of worker complaints or which have a poor safety record, as determined by Coast Guard field office and headquarters review of accident and self-inspection reports, are targeted for unannounced inspections by Coast Guard personnel. These oversight inspections, in turn, provide a means for the Coast Guard to monitor the application and effectiveness of the self-inspection program and to ensure that self-inspections are being conducted properly. A fixed OCS facility owner or operator who files a false self-inspection report is subject to a criminal penalty under the provisions of OCSLA 1978.

DEVELOPMENT OF A NEW PROGRAM

In developing the modified inspection plan, the Coast Guard considered four principal alternatives. The first was to continue the current program in which Coast Guard marine inspectors were responsible for conducting annual inspections of all fixed OCS facilities. This approach was rejected because for some time the Coast Guard had not been able to visit each fixed facility once each year, so there was little incentive for an owner or operator to continually maintain a fixed facility's safety equipment in top condition. Moreover, a court decision holding that OCSLA 1978 requires that the Coast Guard give the owner or operator of a fixed OCS facility notice of a failure to comply with any provisions of the act and a reasonable period for corrective action before it can invoke a penalty made civil penalties an ineffective sanction. The Coast Guard concluded that it would not be able to perform annual inspections of all OCS facilities and follow-up inspections related to compliance without substantially increased staffing and funding dedicated to this purpose.

The second alternative considered was third-party inspection of fixed facilities with owner or operator certification that all deficiencies had been corrected. Under this alternative, an approved third-party would be hired by the owner or operator of a fixed OCS facility. Approved third parties would be designated by the Coast Guard, requiring the establishment of a Coast Guard program to evaluate third-party inspector qualifications. While this alternative would provide a group of qualified, licensed, fixed-facility inspectors in the private sector, the Coast Guard concluded that the attendant administrative burdens to both the industry and the Coast Guard would serve only to decrease efficiency and increase costs with no offsetting improvement in safety beyond that provided by the alternative ultimately selected. Therefore, this alternative was rejected.

The third alternative was to engage MMS personnel to inspect fixed OCS facilities for compliance with Coast Guard regulations. However, involving another federal agency in conducting the inspection that initiates the Coast Guard's enforcement process could lead to unnecessary overlapping of enforcement activities by the two federal agencies–a result proscribed by the OCSLA. Therefore, this alternative was rejected.

The fourth alternative–the one selected–was to require the owner or operator to conduct an annual inspection of the fixed facility utilizing a combination check-off list and reporting form developed by the Coast Guard. The Chief Counsel of the Coast Guard has construed the language of OCSLA 1978 as not mandating that Coast Guard personnel conduct these inspections, but only that the Coast Guard "provide for" the inspections by means of regulations.

---

[3]This change has no effect on the inspection program carried out by Minerals Management Service (MMS) covering, among other items, blowout and pollution prevention.

SUMMARY OF BENEFITS

The self-inspection program is expected to improve safety overall even though unmanned facilities seldom will be visited by Coast Guard inspectors. The Coast Guard, by requiring the industry to conduct the mandated annual inspection, will be able to focus its resources on those fixed OCS facilities that are manned, have a poor safety record, or are the subject of worker complaints. Further, since the Coast Guard will be conducting oversight inspections (spot-checks) of randomly selected manned facilities, many of these facilities will receive multiple inspections during any one year. Additionally, inspection reports involving manned and unmanned facilities and casualty reports now will be reviewed for inconsistencies and analyzed by Coast Guard field units and Headquarters, which will permit the Coast Guard to better evaluate the safety performance of individual operators. It will also provide a mechanism whereby industry trends can be identified or predicted. Fixed facilities with poor safety records (manned and unmanned) will receive additional inspections by Coast Guard marine inspectors. The owners or operators of fixed OCS facilities with good safety records will enjoy the benefit of less government intervention in their operations.

APPENDIX C

OUTER CONTINENTAL SHELF LANDS ACT, EXCERPTS (43 U.S.C.)

§1331. Definition

(b) The term "Secretary" means the Secretary of the Interior, except that with respect to functions under this subchapter transferred to, or vested in, pursuant to the Department of Energy Organization Act (42 U.S.C. 7101 et seq.), the term "Secretary" means the Secretary of Energy, or the Federal Energy Regulatory Commission, as the case may be;

§1332. Congressional declaration of policy

It is hereby declared to be the policy of the United States that -

(1) the subsoil and seabed of the outer Continental Shelf appertain to the United States and are subject to its jurisdiction, control, and power of disposition as provided in this subchapter;
(2) this subchapter shall be construed in such a manner that the character of the waters above the outer Continental Shelf as high seas and the right to navigation and fishing therein shall not be affected;
(3) the outer Continental Shelf is a vital national resource reserve held by the Federal Government for the public, which should be made available for expeditious and orderly development with the maintenance of competition and other national needs;
(4) since exploration, development, and production of the minerals of the outer Continental Shelf will have significant impacts on coastal and noncoastal areas of the coastal States, and on other affected States, and, in recognition of the national interest in the effective management of the marine, coastal, and human environments -

(A) such States and their affected local governments may require assistance in protecting their coastal zones and other affected areas from any temporary or permanent adverse effects of such impacts; and

(B) such States, and through such States, affected local governments, are entitled to an opportunity to participate, to the extent consistent with the national interest, in the policy and planning decisions made by the Federal Government relating to exploration for, and development and production of, minerals of the outer Continental Shelf;

(5) the rights and responsibilities of all States and, where appropriate, local governments, to preserve and protect their marine, human, and coastal environments through such means as

regulation of land, air, and water uses, of safety, and of related development and activity should be considered and recognized; and

(6) operations in the outer Continental Shelf should be conducted in a safe manner by well-trained personnel using technology, precautions, and techniques sufficient to prevent or minimize the likelihood of blowouts, loss of well control, fires, spillages, physical obstruction to other users of the waters or subsoil and seabed, or other occurrences which may cause damage to the environment or to property, or endanger life or health.

(Aug. 7, 1953, ch. 345, sec. 3, 67 Stat. 462; Sept. 18, 1978, Pub. L. 95-372, title II, sec 202, 92 Stat. 634.)

§1348. Oil and gas development and production

(a) Utilization of Federal departments and agencies

The Secretary, the Secretary of the Department in which the Coast Guard is operating, and the Secretary of the Army shall enforce safety and environmental regulations promulgated pursuant to this subchapter. Each such Federal department may by agreement utilize, with or without reimbursement, the services, personnel, or facilities of other Federal departments and agencies for the enforcement of their respective regulations.

(b) Duties of holders of lease or permit

It shall be the duty of any holder of a lease or permit under this subchapter [to]

(1) maintain all places of employment within the lease area or within the area covered by such permit in compliance with occupational safety and health standards and, in addition, free from recognized hazards to employees of the lease holder or permit holder of any contractor or subcontractor operating within such lease area or within the area covered by such permit on the outer Continental Shelf;
(2) maintain all operations with such lease area or within the area covered by such permit in compliance with regulations intended to protect persons, property, and the environment on the outer Continental Shelf; and
(3) allow prompt access, at the site of any operation subject to safety regulations, to any inspector, and to provide such documents and records which are pertinent to occupational or public health, safety, or environmental protection, as may be requested.

(c) Onsite inspection of facilities

The Secretary and the Secretary of the Department in which the Coast Guard is operating shall individually, or jointly if they so agree, promulgate regulations to provide for-

(1) scheduled onsite inspection, at least once a year, of each facility on the outer Continental Shelf which is subject to any environmental or safety regulation promulgated pursuant to this subchapter, which inspection shall include all safety equipment designed to prevent or ameliorate blowouts, fires, spillages, or other major accidents; and
(2) periodic onsite inspection without advance notice to the operator of such facility to assure compliance with such environmental or safety regulations.

(d) Investigation and report on major fires, oil spills, death, or serious injury

(1) The Secretary or the Secretary of the Department in which the Coast Guard is operating shall make an investigation and public report on each major fire and each major oil spillage occurring as a result of operations conducted pursuant to this subchapter, and may, in his discretion, make an investigation and report of lesser oil spillages in one instance of more than two hundred barrels of oil during a period of thirty days. All holders of leases or permits issued or maintained under this subchapter shall cooperate with the appropriate Secretary in the course of any such investigation.

(2) The Secretary or the Secretary of the Department in which the Coast Guard is operating shall make an investigation and public report on any death or serious injury occurring as a result of operations conducted pursuant to this subchapter, and may, in his discretion, make an investigation and report of any injury. For purposes of this subsection, a serious injury is one resulting in substantial impairment of any bodily unit or function. All holders of leases or permits issued or maintained under this subchapter shall cooperate with the appropriate Secretary in the course of any such investigation.

(e) Review of allegations of violations

The Secretary, or, in the case of occupational safety and health, the Secretary of the Department in which the Coast Guard is operating, may review any allegation from any person of the existence of a violation of a safety regulation issued under this subchapter.

(f) Summoning of witnesses and production of evidence

In any investigation conducted pursuant to this section, the Secretary or the Secretary of the Department in which the Coast Guard is operating shall have power to summon witnesses and to require the production of books, papers, documents, or any other evidence shall be compelled by a similar process, as in the district courts of the United States. Such Secretary, or his designee, shall administer all necessary oaths to any witnesses summoned before such investigation.

(g) Report to Congress of violations and action taken

The Secretary shall, after consultation with the Secretary of the Department in which the Coast Guard is operating, include in his annual report to the Congress required by section 1334 of this title the number of violations of safety regulations reported or alleged, any investigations undertaken, the results of such investigations, and the results of the diving studies conducted under section 1347(e) of this title.

(Aug 7, 1953, ch. 345, sec. 22, as added Sept. 18, 1978, Pub. L. 95-372, title II, sec. 208, 92 Stat. 655.)

§1351. Oil and gas development and production

(a) Development and production plans; submission to Secretary; statement of facilities and operation; submission to Governors of affected States and local governments

(1) Prior to development and production pursuant to an oil and gas lease issued after September 18, 1978, in any area of the outer Continental Shelf, other than the Gulf of Mexico, or issued or maintained prior to September 18, 1978, in any area of the outer Continental Shelf, other than the Gulf of Mexico, with respect to which no oil or gas has been discovered in paying quantities prior to September 18, 1978, the lessee shall submit

a development and production plan (hereinafter in this section referred to as a "plan") to the Secretary, for approval pursuant to this section.

(2) A plan shall be accompanied by a statement describing all facilities and operations, other than those on the outer Continental Shelf, proposed by the lessee and known by him (whether or not owned or operated by such lessee) which will be constructed or utilized in the development and production of oil or gas from the lease area, including the location and site of such facilities and operations, the land, labor, material, and energy requirements associated with such facilities and operations, and all environmental and safety safeguards to be implemented.

(3) Except for any privileged or proprietary information (as such term is defined in regulations issued by the Secretary), the Secretary, within ten days after receipt of a plan and statement, shall (A) submit such plan and statement to the Governor of any affected State, and, upon request to the executive of any affected local government, and (B) make such plan and statement available to any appropriate interstate regional entity and the public.

(c) Scope and contents of plan

A plan may apply to more than one oil and gas lease, and shall set forth, in the degree of detail established by regulations issued by the Secretary -

(1) the specific work to be performed;
(2) a description of all facilities and operations located on the outer Continental Shelf which are proposed by the lessee or known by him (whether or not owned or operated by such lessee) to be directly related to the proposed development, including the location and size of such facilities and operations, and the land, labor, material, and energy requirements associated with such facilities and operations;
(3) the environment safeguards to be implemented on the outer Continental Shelf and how such safeguards are to be implemented;
(4) all safety standards to be met and how such standards are to be met;
(5) an expected rate of development and production and a time schedule for performance; and
(6) such other relevant information as the Secretary may by regulation require.

(j) Cancellation of lease on failure to submit plan or comply with approved plan

Whenever the owner of any lease fails to submit a plan in accordance with regulations issued under this section, or fails to comply with an approved plan, the lease may be canceled in accordance with section 1334(c) and (d) of this title. Termination of a lease because of failure to comply with an approved plan, including required modifications or revisions, shall not entitle a lessee to any compensation.

APPENDIX D

DATA ON PINCs AND INCs

- Excerpts from MMS Report 3704 series for two operators (A and B). INC/PINC Ratio by Operator-Production
- Excerpt from MMS Report 3702 series, INC/PINC Ratio by PINC Number-Production

Report 3704
Excerpt: INC/PINC Ratio by Operator-Production
Company A
* W - Warning, S - Shut-in, C - Component shut-in

PINC number	Ratio	Times checked	Yes	No	NA	Action*
P178	0.000	3	3	0	0	W
P178	0.000	37	37	0	0	
P179	0.000	26	26	0	0	
P180	0.000	25	25	0	0	
P181	0.000	23	23	0	0	
P200	0.000	1	1	0	0	C
P200	0.000	97	97	0	0	
P201	0.000	49	49	0	0	
P202	0.000	69	69	0	0	
P203	0.000	42	42	0	0	
P204	0.000	15	15	0	0	
P205	0.000	33	33	0	0	
P206	0.000	17	17	0	0	
P207	0.000	11	11	0	0	
P250	0.000	31	31	0	0	
P251	0.000	35	35	0	0	
P252	0.000	4	4	0	0	W
P252	0.000	36	36	0	0	
P253	0.000	17	17	0	0	
P255	0.000	34	34	0	0	
P300	0.000	53	53	0	0	
P300	1.000	1	0	1	0	C
P301	0.000	111	111	0	0	
P350	0.000	6	6	0	0	W
P350	0.000	303	303	0	0	
P351	0.000	19	12	0	7	
P352	0.000	145	111	0	34	
P352	0.104	48	43	5	0	C
P400	0.000	176	142	0	34	
P400	0.111	18	16	2	0	C
P401	0.000	165	131	0	34	
P401	0.069	29	27	2	0	C
P402	0.000	107	87	0	20	
P402	0.100	84	63	7	14	C
P403	0.000	37	15	0	22	
P450	0.000	5	5	0	0	
P451	0.000	5	5	0	0	
P452	0.000	5	5	0	0	
P500	0.000	39	39	0	0	
P501	0.000	39	39	0	0	
P502	0.000	37	37	0	0	
P503	0.000	54	54	0	0	
P504	0.000	39	39	0	0	
P505	0.000	109	68	0	41	

Report 3704
Excerpt: INC/PINC Ratio by Operator-Production Company B
* W - Warning, S - Shut-in, C - Component shut-in

PINC number	Ratio	Times checked	Yes	No	NA	Action*
E100	0.000	1	1	0	0	
E101	0.000	1	1	0	0	
E102	0.000	4	4	0	0	
E103	0.000	1	1	0	0	
E104	0.000	8	8	0	0	
G100	0.000	7	7	0	0	
G100	1.000	1	0	1	0	W
G101	0.000	44	44	0	0	
G102	0.000	8	8	0	0	
G200	0.000	8	8	0	0	
G300	0.000	1	1	0	0	
G301	0.000	3	2	0	1	
G400	0.000	2	2	0	0	W
G400	0.000	6	6	0	0	
G500	0.000	7	7	0	0	
G501	0.000	7	7	0	0	
G502	0.000	7	7	0	0	
G503	0.000	7	7	0	0	
P100	0.000	4	4	0	0	
P101	0.000	4	4	0	0	
P102	0.000	1	1	0	0	
P104	0.000	3	3	0	0	
P105	0.000	3	3	0	0	
P106	0.000	1	1	0	0	
P107	0.000	1	1	0	0	
P108	0.000	4	4	0	0	
P109	0.000	4	4	0	0	
P110	0.000	1	1	0	0	
P111	0.000	1	1	0	0	
P150	0.000	3	3	0	0	C
P150	0.000	44	44	0	0	
P151	0.000	20	20	0	0	
P153	0.000	28	28	0	0	
P154	0.000	3	3	0	0	
P159	0.000	43	43	0	0	
P160	0.000	2	2	0	0	
P161	0.000	7	7	0	0	
P162	0.000	3	3	0	0	
P163	0.000	3	3	0	0	
P164	0.000	3	3	0	0	
P165	0.000	3	3	0	0	
P166	0.000	1	1	0	0	
P167	0.000	1	1	0	0	

MINERALS MANAGEMENT SERVICE
OFFSHORE INSPECTION SYSTEM
INC/PINC RATIO BY PINC NUMBER

OCSIS ID: 3702 DATE: 08/31/89

PINC NUMBER	RATIO	TIMES CHECKED	YES	NO	NA	ENF* ACTION	DESCRIPTION
P207	0.000	138	138	0	0		TSE used in heat exchangers as required by API RP 14C7
P207	0.008	143	125	1	17	C	TSE used in heat exchangers as required by API RP 14C7
P250	0.000	459	459	0	0		are operable manually-operated quick-openings ESD valves located at heli
P250	0.000	692	685	0	7	W	are operable manually-operated quick-openings ESD valves located at heli
P250	0.000	9	9	0	0	S	are operable manually-operated quick-openings ESD valves located at heli
P250	0.000	7	7	0	0	C	are operable manually-operated quick-openings ESD valves located at heli
P251	0.000	655	655	0	0		are operable manually-operated quick-openings ESD valves located at exit
P251	0.002	839	834	2	3	W	are operable manually-operated quick-openings ESD valves located at exit
P251	0.000	6	6	0	0	S	are operable manually-operated quick-openings ESD valves located at exit
P251	0.000	7	7	0	0	C	are operable manually-operated quick-openings ESD valves located at exit
P252	0.000	669	669	0	0		are operable manually-operated quick-openings ESD valves located at boat
P252	0.010	975	965	10	0	W	are operable manually-operated quick-openings ESD valves located at boat
P252	0.000	11	11	0	0	S	are operable manually-operated quick-openings ESD valves located at boat
P252	0.000	7	7	0	0	C	are operable manually-operated quick-openings ESD valves located at boat
P253	0.000	139	139	0	0		are operable manually-operated quick-openings ESD valves located at cent
P253	0.005	212	192	1	19	W	are operable manually-operated quick-openings ESD valves located at cent
P253	0.000	2	0	0	2	S	are operable manually-operated quick-openings ESD valves located at cent
P253	0.000	1	1	0	0	C	are operable manually-operated quick-openings ESD valves located at cent
P254	0.000	14	14	0	0		are operable manually-operated quick-openings ESD valves location near t
P254	0.000	57	38	0	19	W	are operable manually-operated quick-openings ESD valves location near t
P254	0.000	1	0	0	1	S	are operable manually-operated quick-openings ESD valves location near t
P255	0.000	556	537	0	19		are operable manually-operated quick-openings ESD valves location near t
P255	0.006	548	508	3	37	W	are operable manually-operated quick-openings ESD valves location near t
P255	0.000	3	3	0	0	S	are operable manually-operated quick-openings ESD valves location near t
P255	0.000	4	3	0	1	C	are operable manually-operated quick-openings ESD valves location near t
P300	0.000	938	937	0	1		exhaust piping from each diesel engine equipped with spark arrestors?
P300	0.002	911	897	2	12	C	exhaust piping from each diesel engine equipped with spark arrestors?
P300	0.000	5	4	0	1	W	exhaust piping from each diesel engine equipped with spark arrestors?
P301	0.001	1686	1672	1	13		each engine exhaust insulated and piped away from the fuel source acc
P301	0.003	1719	1697	5	17	C	each engine exhaust insulated and piped away from the fuel source acc
P301	0.000	17	17	0	0	W	each engine exhaust insulated and piped away from the fuel source acc
P350	0.000	3034	2998	0	36		does each wellhead have a rated WP greater than the surface shut-in pres
P350	0.000	4487	4445	0	42	W	does each wellhead have a rated WP greater than the surface shut-in pres
P350	0.000	9	9	0	0	C	does each wellhead have a rated WP greater than the surface shut-in pres
P351	0.000	211	147	0	64		are two master valves installed if the surface pressure is in excess of
P351	0.000	600	541	0	59	C	are two master valves installed if the surface pressure is in excess of
P351	0.000	6	6	0	0	W	are two master valves installed if the surface pressure is in excess of
P352	0.002	2450	1921	4	525		an operable SSV installed on the second valve in the flow stream from
P352	0.034	3838	3387	119	332	C	an operable SSV installed on the second valve in the flow stream from
P352	0.000	20	20	0	0	W	an operable SSV installed on the second valve in the flow stream from
P400	0.002	2736	1974	4	758		each flow line segment equipped with a PSH in accordance with API RP
P400	0.021	3689	3284	72	333	C	each flow line segment equipped with a PSH in accordance with API RP
P400	0.000	5	5	0	0	W	each flow line segment equipped with a PSH in accordance with API RP
P401	0.001	3013	2319	3	691		each flow line segment equipped with a PSL in accordance with API RP

* W - Warning, S - Shut-in, C - Component shut-in

APPENDIX E

MMS INSPECTOR POSITION DESCRIPTION: PETROLEUM ENGINEERING TECHNICIAN GS-0802-11 OFFSHORE OIL AND GAS DISTRICT

MAJOR DUTIES

Conducts onsite inspections of complex offshore oil and gas drilling and workover rigs, including new rigs, to assure that the operator's equipment, safety systems, and operations are in compliance with approved plans. Federal regulations, OCS Orders, and related standards and procedures, and to provide for protection of the human, marine, and coastal environments. Independently, or with the assistance of lower-graded technicians, carries out the following duties:

- regularly serves as inspection team leader with responsibility for advising, guiding and training lower-graded technicians on drilling and workover safety requirements;
- as assigned, serves as leader or member of accident investigation teams appointed to investigate blowouts, serious accidents, and other disastrous events;
- independently orders the shut-in of components causing pollution; independently selects optional enforcement codes, and when the codes permit, independently orders the shut-in of drilling and workover operations, or components or operations affected; when the enforcement codes do not permit independent action, obtains supervisory approval prior to ordering the shut-in of drilling and workover operations, and components; independently authorizes the resumption of operations when deficiencies are corrected prior to leaving the rig;
- checks operator field records and logs to assure that required information is properly obtained and recorded and that operations are in accordance with approved applications and other regulatory and procedural requirements;
- observes working conditions, procedures, and practices to assure that they are proper and safe;
- inspects and verifies proper placement and condition of safety equipment;
- inspects or witnesses all types of drilling workover equipment and operations, including running and cementing of casing, pressure testing, cutting and recovering casing, directional surveys, logging and plugging and abandonment of wells, pollution-control equipment, mud-system monitoring, installation and testing of blow-out prevention equipment including sophisticated subsea, well testing, workover and wireline operations, well-control surveillance, well completions, installation and testing of H_2S and gas detection devices;
- prepares inspection reports;
- conducts follow-up inspections to assure that proper corrective actions have been taken;

- performs office work related to drilling and workover operations and inspection activities, including letters of deficiencies to operators;
- is occasionally required to review operators' drilling and workover plans and applications, including schematics and diagrams, and recommends approval or modifications, as appropriate;
- is occasionally required to conduct initial inspections of a new rig and rigs new to the area, to assure that all required safety and pollution-prevention equipment has been installed; when deficiencies are found, orally notifies operator of corrective action required prior to commencement of drilling operations;
- operates Government vehicles as an incidental driver;
- performs other incidental or related duties as assigned.

1. Knowledge Required by the Position

This position requires a thorough and in-depth knowledge of well drilling and workover systems, processes, methods, and equipment utilized by the petroleum industry in the offshore exploration and development of oil and gas resources. This includes knowledge of (1) the function, capabilities, and limitations of the various items of equipment involved in drilling and workover operations; (2) oil and gas flow, temperatures, and pressures, and their significance in the drilling and workover processes; (3) well casing and cementing operations, and drill-stem testing procedures; (4) the various types of blowout preventers and related well-control equipment; (5) characteristics and special hazards of H_2S and SO_2; and (6) the safety principles and practices related thereto.

As regular team leader, inspecting all types of drilling workover equipment and operations, the full range of this knowledge is used to independently: (1) Assess the potential safety impact on the human, marine, and coastal environments by the systems, processes, and equipment utilized on drilling and workover rigs. (2) Review drilling and workover equipment schematics and diagrams and operator applications and reports. (3) Effectively inspect operations and thoroughly evaluate safety situations, including those not covered by existing regulations and procedures, or involving borderline incidents of noncompliance. (4) Assure compliance with operational procedures. As regular team leader, the knowledge is used to guide, advise, train, and develop the capabilities of lower-grade technicians, both in the field and in the office.

Uses general working knowledge of production operations to occasionally assist in performing inspections of such operations as peak workloads may require and for cross-training purposes.

2. Supervisory Controls

Serving regularly as inspection team leader: (1) The technician's supervisor schedules and assigns inspections. (2) Independently plans and performs all types of inspections, adapting the inspection approach and the procedures used to actual conditions on the facility; takes independent action on situation and circumstances not adequately covered by guidelines, procedures, and precedents; independently selects the most appropriate action when guidelines provide for optional enforcement actions; discusses deficiencies of pollution-prevention devices, safety equipment, and associated practices with responsible onsite operator personnel; provides advice to operator personnel on ways to best achieve compliance; orders the shut-in of operations while on the facility, either independently or with supervisory approval, in accordance with guidelines on enforcement actions and responsibilities; conducts follow-up inspections while on-site, independently authorizes resumption of operations when deficiencies are corrected prior to leaving the site; independently prepares written inspection reports; independently prepares written warnings and shut-in notices for on-site delivery to operator personnel; prepares letters to operator supervisory representatives in final form concerning corrective actions needed for compliance, and forwards through channels for

District Supervisor's signature. (3) The technician's work is occasionally reviewed for adherence to operating regulations, inspection policies, and guidelines.

3. Guidelines

Written guidelines include approved drilling and workover programs; Federal regulations; OCS Orders; field rules; Division and other instructional memoranda; safety-alert notices and other operational issuances; standards and codes of the petroleum and related industries (e.g., American Petroleum Institute, National Electric Code; and National Fire Protection Association); and drilling workover, and production inspection forms which include more than 400 potential incidents of noncompliance. Occasional oral guidelines are provided by the supervisor.

The technician regularly exercises independent technical judgment in interpreting standards, regulations, Orders, and procedures, and when they provide for alternatives, are not clearly applicable, or do not specifically address the situation, determines appropriate courses of action to be taken.

4. Complexity

Drilling and workover inspections involve routine to highly sophisticated equipment and operations characterized by the following:

- facilities ranging from those of moderate size and complexity to the newest, largest, and most complex offshore facilities used by the petroleum industry. They include ocean-going drill ships, submersibles, semi-submersibles, jack-ups, and fixed platform facilities;
- numerous and varied types of well control and other rig safety devices, including annular and ram-type blowout preventers, diverter systems with automatic remote-controlled valves, mud-system monitoring equipment, pressure valves and gauges, chokes, locking devices and other surface and subsurface safety devices, and sophisticated and complex electronic/computerized monitoring equipment;
- elaborate equipment systems and procedures required for drilling and safety control, including electrical systems for power supply and backup auxiliary supply, gas-detection systems and alarms, pollution-prevention systems, and procedures and equipment for protection, survival, and escape of personnel;
- a substantial amount and wide variety of heavy-equipment items and drilling materials, including derrick and other hoisting equipment, winches, motors, generators, tanks drill pipe, casing, cement, drilling mud, and chemicals;
- extensive support facilities often required to quarter and feed operator and contractor personnel;
- repair, construction, drilling, workover, and wireline activities frequently performed on the platforms simultaneously with production operations, thereby increasing the danger of hazards to personnel and the environment.

The technician performs inspections of all types of drilling and workover equipment and operations, including field records, working conditions and practices, running and cementing of casing, pressure testing, cutting and recovering casing, directional surveys, logging, plugging and abandonment of wells, pollution-control equipment, proper placement and condition of equipment; mud-system monitoring, installation and testing of blowout-prevention equipment including sophisticated subsea equipment, well testing, workover and wireline operations, well-control surveillance, well completions, and installation and testing of H_2S and gas detection devices.

The inspection of drilling and workover operations is critical from the standpoint of personnel and environmental safety. When a kick or blowout occurs during an inspection, the drilling technician monitors all operations until the kick or blowout is controlled. Moreover, when drilling involves new or relatively unexplored areas, the complexity of the inspection is further increased since the technician must devote special attention to the possibility of abnormal pressures, kicks, and blowouts. Similarly, when an H_2S environment is known, expected or encountered, complexity is further increased since the technician must devote particular attention to special preventive measures, equipment, and procedures necessary to control the effect of the toxicity, flammability, and corrosive characteristics of H_2S. Also, during the initial inspection of a new rig, the complexity of the inspection or modification in standard operating practices, equipment, safety devices, and system designs.

As an accident-investigation team leader or member, the technician contributes his practical and comprehensive knowledge of accepted operating practices and related safety requirements to the team's efforts. Accidents and disastrous events investigated are complex because the causes, or probable causes, are often not readily identifiable and require in-depth technical analysis of the conditions and circumstances surrounding the event.

The technician is required to review and interpret complex drawings, specifications, and performance data of equipment, programs, safety systems, and pollution-prevention equipment associated with the drilling or workover operations being inspected. The technician also is required to review and interpret complex OCS Orders and other regulatory safety requirements.

The occasional inspection of oil and gas production operations, in addition to the regular inspection of drilling and workover operations, serves to further broaden and increase the complexity of the technician's duties. While these latter duties are assigned only occasionally, the technician must be fully capable of assisting in conducting them when inspection workload peaks or other circumstances require such assignments.

As regular team leader, the technician is often required to analyze difficult, innovative, and complex technical situations, and to exercise originality and independence in determining appropriate action.

5. Scope and Effect

The scope of the work includes enforcing and promoting pollution prevention and efficient and safe operations among the many complex offshore drilling and workover facilities for the protection of the human, marine, and coastal environments. This is accomplished by (1) independently conducting onsite inspections of the processes and equipment and practices involved, and reporting deficiencies to operators and assuring corrective action; (2) leading or participating in the investigation of serious accidents and disastrous events, identifying causes and measures to prevent recurrences; (3) reviewing and advising on the adequacy of safety systems as a result of onsite inspections; (4) as inspection team leader, providing advice, guidance, and training to lower-graded technicians; (5) making recommendations pertaining to the issuance of new or revised OCS Orders and other regulatory requirements; and (6) occasionally assisting in the inspection of nearly all aspects of production operations.

The effect or result of these activities is the identification and correction of safety hazards for the prevention of accidents and disastrous events. Onsite inspection and evaluation of drilling and workover operations have a favorable impact on the development of new safety systems and practices by the offshore petroleum industry. When safety hazards pose an immediate danger, operations are shut-in until deficiencies are corrected. The thoroughness and quality of the technician's work have substantial impact on the agency program in safeguarding and furthering the exploration, development, and utilization of the Nation's offshore oil and gas resources. Future development of oil and gas resources in frontier Outer Continental Shelf areas is dependent in large measure upon the continued effectiveness of the agency program in preventing major oil spills and other polluting events in mature areas. Through the activities described above, the technician plays an important role in achieving this National objective.

While secondary in importance to the safeguarding of the human, marine, and coastal environments, decisions made by the technician also have substantial economic impact. When operations are shut in, development of critically needed domestic oil and gas is delayed. Such delays can cost the operator many thousand of dollars a day (rigs are leased to operators for amounts up to $40,000 daily, and the cost of leasing and operating a large rig can exceed $100,000 a day).

6. Personnel Contacts

Personnel contacts are typically with helicopter pilots; boat operators; other technicians, radio operators, professional engineers, geologists, and geophysicists in the agency; and operator and contractor supervisory officials and other representatives.

7. Purpose of Contacts

The purpose of contacts with petroleum industry personnel is to obtain compliance with Federal regulations, OCS Orders, and related safety standards and procedures designed to facilitate pollution prevention and the efficient and safe operation of offshore facilities. During these contacts, the technician regularly deals with different operators, contractors, supervisory attitudes, and routine to complicated and controversial safety issues. The technician uses tact and persuasion, along with technical expertise and authority, to respond to challenges to his decisions and to obtain quick and effective correction of safety deficiencies.

The purpose of contacts with agency personnel is to receive guidance, provide and exchange information, and discuss job-related matters. The purpose of contacts with lower-graded technicians is to further their development and effectiveness as technicians.

8. Physical Demands

Onsite inspection work requires almost continuous physical activity. Most of the time is spent walking, standing, crouching, and climbing. The technician must walk under and around heavy derrick and hoisting equipment and piping, and climb up and down ladders and steep stairways between floors and decks.

9. Work Environment

The technician works around potentially hazardous operating equipment. In climbing up and down ladders and steep stairways, the technician is protected from falls only by a firm grip on ladder rungs and handrails. The work environment involves exposure to high-frequency noise, extreme weather conditions, explosions, fires, poisonous gases, toxic chemicals, high elevations over water, and high-pressure equipment. Transportation to and from the work site involves helicopter flying over water, and occasional use of small boats in rough seas. The work environment requires the use of hard hats, hard-toe boots, safety glasses, ear protection equipment, life vests or survival suits, flight safety helmets, cold-weather gear, and protective breathing gear.

APPENDIX F

COMMENTARY ON REPRESENTATIVE SAFETY GOALS AND APPROACHES

The question of "how safe is safe enough," is one that various government agencies have addressed within the context of their mission. The science of risk assessment and management to address these issues has advanced greatly in recent years.

MODERN SAFETY GOALS

Public Safety

There is an established but still evolving discipline for determining acceptable levels of public safety. Chauncey Starr, of the Electric Power Research Institute (EPRI), published seminal articles on this topic in the 1970s. A large volume of literature and analyses has developed since then, holding that it is especially vital for a public agency to have an explicit and defensible rationale for its safety goals. In the absence of explicit, rational goals, systems become vulnerable to attack from a "zero-risk" viewpoint, which has perennial appeal as a political posture. Such situations have developed in some areas of the OCS where leasing is being considered. When these situations develop, any deviation from perfection in operation even if it does not result in fatalities, serious injuries, or property damage can be portrayed as a "near-miss" (or more precisely, a near-hit) to a disaster.

Environmental Safety

Acceptable environmental safety is even more complex an issue than acceptable public safety, since esthetic and fundamental philosophical considerations are involved as well as objective measures of actual and potential damage. As to public safety, actual practice shows that activities contributing less than 0.1 percent of the total average yearly risk of death or disability (less than 1 in 1,000) rarely are subject to intensive activity or regulation to reduce the risks. Considerably higher risks are tolerated for some activities. For environmental hazards, however, the measures of damage are more subjective and value-laden. An acceptable algebra of incommensurables is yet to be developed. For example, how does one scale the potential loss of a particular subpopulation of snail-darters against the occasional occurrence of tar or oil on a beach, or the visible presence of a distant steel structure on a seascape? The objective measure of potential damage from spills or blowouts usually is tempered by the relatively short duration of severe and demonstrable consequences. However, the perception that any event is a precursor to a much greater disaster is not uncommon. This can often result in demands for extension of jurisdiction by federal and state agencies, greater activism in regulation, intensified deterrence or prohibitions of industrial and commercial activities, and litigation.

The implication of these perceptions is that it is prudent for both government and industry to provide state-of-the-art levels of safety management and protection against severe accidents of low probability, and also against less severe "environmental insults" than those that historically have been accepted.

SPECIFIC EXAMPLES

Nuclear

The Nuclear Regulatory Commission (NRC) uses a highly developed discipline for estimating the likelihood of severe events of low probability. The importance of this discipline is that it provides a systematic and cumulative record of all of the hazards that have been experienced or that can be imagined (within the bounds of physically possible events.)

This discipline, Probabilistic Risk Analysis (PRA), also provides a method for estimating the likelihood of highly improbable events that may never have happened, but that are conceivable and physically possible. The method involves the systematic tabulation of the different sequences that can occur when things fail, break, leak, or are subject to errors in control functions or human performance.

Initially, the method depended largely on failure rate data from fossil-fired power plants and chemical plants. (For example, for control systems, pipes, valves, motors, vessels, etc.) Since the original study (WASH-1400, 1975) the data from over 3,000 unit-years of nuclear plant operation have been accumulated. Since 1979, these data have been systematically recorded and analyzed for lessons learned and for remedial measures necessary to reduce the likelihood of recurrence. Three computerized data bases capture component reliability data, along with the events data that also included human error as a frequent element.[1] These data bases provide a sound basis for estimating and managing the risk levels for any conceivable type of event, to about the level of 1 in 10,000 likelihood per year, per plant. While these data bases go well beyond the needs of the OCS industry and MMS, they do illustrate the process for acquiring data to conduct PRAs (Probabilistic Risk Analyses) and establish one precedent for having a common, accessible data base provided from industry sources.

Occupational Safety and Health Administration (OSHA)

OSHA now sets standards for occupational exposures on the basis of a systematic risk assessment discipline. The intent is to define allowable levels in terms of a non-zero expected response (toxic, carcinogenic, mutagenic or teratogenic) of an average population. The goals usually are in the range of one expected effect per ten thousand to one million exposed individuals. The target range is large because of the inherent uncertainties in extrapolation of animal data to

---

[1]The three data bases are *License Event Report*, which is NRC recorded onsite and which deals with lessons learned for any nonstandard events, including safety-related ones, as defined by the operating license; the *Nuclear Plant Reliability Data System*, which is a report maintained by the Institute for Nuclear Power Operations (INPO), based on data provided by the operator and covering component-related performance and quality history; and the *Significant Operating Experience* report, which is a "must fix" report from the industry audits conducted by INPO. A large bulk of this information is maintained by INPO and is open to all utilities; this information base includes input from 14 foreign countries.

humans, in the actual net levels delivered, in variations in individual response, and in the different effects of short-term versus chronic exposures.

Food and Drug Administration (FDA)

The FDA increasingly has used systematic risk assessment. This trend was driven in part by the dilemmas posed by the Delaney Amendment, which banned the use of any material that was a known or suspected carcinogen in animal tests. This zero-risk approach involved serious absurdities once it was recognized that nearly all classes of foodstuffs, in a natural unadulterated condition, contain substances that are carcinogenic in animal tests.

The risk assessment methodology has provided a defensible basis for establishing that sufficiently small amounts of such substances are acceptable and probably necessary for life itself. This has been reasonably successful in countering the zero-risk posturing that sometimes drives legislatures.

APPENDIX G

COMMENTARY ON THE FEASIBILITY OF SAMPLING

Several of the alternative inspection programs evaluated by the committee (alternatives 2 and 3, in particular) are based on the premise that sampling procedures can be used under which–within each category of equipment–only a representative sample of the items need to be tested. Thus, for instance, on a given platform with 100 high-level sensors, the current inspection practice requires that all 100 sensors be tested while the inspector observes. The alternative programs are based on the proposition that with statistical sampling techniques the inspector need witness only a limited number of these 100 sensors to establish the reliability of all the sensors.

The result of this approach is to free inspectors' time, which could then be put to use making more unannounced inspections, following up on weak spots, and carrying out essential tasks such as those described in Chapter 7. These are perceived by the committee as more effective uses of the inspectors' time than repetitive witnessing of routine tests.

Under the alternative inspection programs the platform operator would continue to be responsible for performing all the routine tests, on the schedule MMS has prescribed; the thing that would be new is that now he would be required on demand to send records of any of these test results to the MMS (probably, he also would have to accommodate an MMS prescribed report format). The nature of MMS inspections, however would change; it now would verify that the operator was performing the tests and that his reported results are accurate, rather than witnessing all the annual tests as they are conducted by the operator. The purpose of sampling then, would be to verify operator performance of mandated tests.

Such verification can be done using "statistical inference" methods. For example, in his report to MMS the operator states that in testing for the past year the failure rate of High Level Switches (LSHs) was 0.01; however, MMS is receiving reports from all operators and is maintaining a data base from which MMS knows that the industry average failure rate for LSHs is 0.02. Moreover, MMS knows that the reported number falls in the top (best) fifth percentile of all platforms. MMS would now ask "does the operator's record in past inspections, his overall safety performance, and everything else we know about him support the assertion that now he is among the top five percent performers?" If not, MMS would plan to witness more tests during its onsite inspection than it would normally witness, i.e., it would establish a somewhat larger sample size for his platform.

Suppose the MMS inspector, now on the platform, witnesses tests of a sample of 10 LSHs, chosen at random but with preference toward the "hard to get to" units, and one out of these 10 fails the test. Referring to statistical tables prepared on the basis of MMS data, the inspector sees that he has reason to suspect that the operator's reported data are not accurate.[1] Again referring to statistical tables, he determines that further verificaton of the operations performance is indicated, and orders 20 more tests, during which he finds 2 additional failures. He now orders a 100 percent inspection of LSHs and all other items on the PINC list. If the operator is found to have been falsifying records, appropriate penalties would be applied, ranging from more frequent unannounced inspections to shut-ins, to criminal charges.

The committee believes that it is feasible to develop valid sampling schedules for the industry with inspection increments based on the comparability of the operator's reports with his operating history and those of other operators, as well as test failures encountered in the course of the onsite inspection.

With limited retraining current MMS inspectors can be trained to use sampling schedules. A changeover to a sampling procedure cannot and should not, however, be executed overnight. The sampling procedure should be phased in gradually; moreover, the preparation and initial implementation of sampling procedures will require the services of persons skilled in this kind of statistical inference.

---

[1]Certain failures, such as out-of-tolerance readings, do occur routinely. If an MMS inspector examines an operator's records, he would expect to see failures and corrective action recorded at about the statistical average for the industry. If he does not, he would look for problems with the operator's testing, recordkeeping, and/or reporting practices.

APPENDIX H

COMMENTARY ON UNITED KINGDOM AND NORWEGIAN INSPECTION REQUIREMENTS[1]

The committee reviewed the rules and regulations governing United Kingdom (U.K.) and Norwegian inspection of drilling and production operations. These cannot be related directly to the U.S. outer continental shelf (OCS) requirements in the Gulf of Mexico and the Pacific Coast because of differences in environmental conditions in the different areas. North Sea operations typically utilize large, enclosed structures with a high concentration of personnel in an environment characterized by severe cold weather, high waves, and cold seas. It is not uncommon for a North Sea installation to have over 200 personnel and to contain wells, drilling equipment, producing equipment, and quarters. There are long periods of time when safe evacuation of platforms is not possible and survival of individuals who might be cast into the cold water would be highly uncertain.

U.S. OCS operations typically utilize structures of open construction (being thus well ventilated) which are much smaller than North Sea structures and contain much smaller quarters (up to 75 people). The environment is relatively benign. There are only short periods of time when evacuation is not possible, and abandonment by jumping into the water is feasible with an expected high survival rate.

Although the safety risks being addressed are much different, it is instructive to review U.K. and Norwegian inspection philosophies. While both countries are faced with similar situations they have elected different approaches to operational safety inspection due, in part, to their societal differences as well as to the manner in which the offshore industry developed in the two countries.

UNITED KINGDOM

In the U.K. system, independent certifying agencies approved by the Department of Energy carry out most of the certification and inspection functions. The certifying agencies (1) review the original design and construction; (2) check calculations, quality control procedures, and documentation to ensure that design and construction conform to both the standards of safety required by the Department of Energy and those required by the certifying agency; and (3) issue certificates of fitness. The certifying agencies perform inspections at least annually to ensure that the structure and equipment are maintained and operated in accordance with these standards.

This philosophy parallels that applied by the U.K. to its maritime industry. Indeed, five of the six approved certifying agencies are the historic "classification societies" used to "class" ships (i.e., ensure they meet safety standards) for insurance purposes. These are American Bureau of Shipping, Bureau Veritas, Det norske Veritas, Germanisher Lloyd, and Lloyds Register of Shipping. Halcrow Eubank and Associates Certification Group is also approved by the Department of Energy.

---

[1]Based on requirements in force as of September 1987.

It is the responsibility of the operators to contract with one of these organizations to provide the necessary initial "Certificate of Fitness" and the necessary annual renewals. The regulations state, "The owner has a statutory responsibility for the sound design, proper construction, and effective maintenance of his installation, including adequate supervision and inspection necessary to achieve these ends, but a valid Certificate of Fitness is evidence that an independent and responsible organization believes the owner to have honored his statutory obligations."

The Department of Energy issues Guidance Notes to aid certifying authorities in ensuring compliance. Some of these (particularly those pertaining to structural integrity) are very specific, while others are much broader in scope, allowing room for interpretation by the certifying authority. An example of a broad statement is the requirement that the installation equipment be reviewed to assess that "there are included those items which are required to be incorporated, that all necessary safety features are incorporated and that all equipment is in accordance with recognized standards." Examples of more specific instructions which still allow considerable leeway are as follows:

- Each pressure vessel is required to be examined by the certifying authority "externally and internally and tested as appropriate at intervals of time to be determined by the certifying authority."
- "It is the responsibility of the owner to perform regular tests and inspection of the ESD system . . . to be agreed between the owner and the Certifying Authority."

An example of highly specific guidance is that which is provided for electrical system inspection. A brief excerpt is set out below:

2.2.1 Annual surveys

The recommendations contained in the IEE RECOMMENDATIONS and the D.En. CODE should be carried out.

A general examination should be carried out of the electrical installation (generators, switchgear, cables, cable glands, machinery, lighting, etc.).

Without prejudice to the generality of the above the following should be carried out during a subsequent survey:

2.2.1.1 Cable runs should be examined for sheath or armor defects and to ensure that the means for supporting the cable is in good order.

2.2.1.2 Sample checks should be made of equipotential bonding conductors to verify the earth continuity path.

2.2.1.3 Electrical equipment located in hazardous areas should be examined to ensure that it is suitable for the application and that the integrity of the protection concept has not been impaired.

Alarms and interlocks associated with pressurized equipment and spaces should be tested to ensure correct operation.

2.2.1.4 The emergency source of power, its associated circuits and, where fitted, the transitional emergency source of power should be tested.

Even more detail is included for inspection and documentation of material and fabrication of structural members, and drilling and production components. For example:

3.6.4 Material Documentation and Identification

Major load bearing and pressure retaining parts of drill through equipment should be manufactured with supporting documentation that ensures traceability through the manufacturing process. Components requiring traceability should be identified by lot number or individual serial number to maintain traceability throughout manufacture. The component parts should be identifiable within the assembly. The owner should ensure that the manufacturer is responsible for maintaining this documentation on file and upon request for providing this information to the owner for his retention, for a period of not less than 5 years after shipment. Traceability documentation may include, but is not necessarily limited to the following:

A. Material source report
   - Chemical and mechanical properties for each heat
   - Heat treatment temperatures and time at temperature
   - Charpy temperature and impact values
   - Hardness test readings (verification to NACE MR-01-75)

B. Manufacturing processes
   - Welding
   - Post weld heat treatment
   - Hardness test results (verification to NACE MR-01-75)
   - Hydrostatic pressures tests
   - Dimensional check results

3.6.5 Documentation of Component Assemblies

The manufacturer of assembled components should document the supporting information as related to the purchaser's requirements and maintain the same for his file.

The owner may request the manufacturer to prepare a manufacturing report consolidating the documentation into a "data book" for retention by the manufacturer. Specific data may be issued to the owner upon request.

The manufacturing report should include the following documentation as appropriate:

A. Purchase order number/sales order number
B. Basic design criteria
C. Assembly arrangement drawing with material identification and where used
D. Manufacturer's statement of compliance
E. Material documents - see Sub-Section 3.3
F. Welding procedure qualification
G. Welder and operator qualification
H. NDE operator qualification
I. NDE results - see Sub-Section 3.6.3
J. Assembly and pressure test report (including certified charts)
K. As built overall dimensions and actual weights
L. Independent design review document

Although addenda to the Guidance Notes tend to add more detail (perhaps as specific problems are identified), there still is considerable leeway for the certifying authority to determine items to be inspected, inspection frequencies, and standards for inspection. The Department of Energy audits the certifying authorities to ensure that they are performing their responsibility in a competent and consistent manner. However, operators report that differences exist. It is said to be common, in inviting bids from certifying authorities to perform the work for a certificate of fitness, for the operator to receive a cost spread of 3 to 1 in the bids. This could indicate that the scope of the inspection effort may vary significantly among the approved certifying authorities.

The nature and amount of effort expended by the Department of Energy in auditing compliance of certifying agencies was not determined by the committee.

NORWAY

The Norwegian Petroleum Directorate (NPD) uses it own personnel to monitor safety and requires that operators set up and comply with an approved internal control plan. The Norwegian regulations are more detailed than U.K. regulations in regard to the overall work environment (e.g., standards of hygiene, medical equipment, work rules, quarters requirements, etc.). The operator is required to develop and follow detailed safety instructions, contingency plans, and safety drills. A Safety Delegate Committee and a Work Environment Committee are required, with membership (which includes employees) and authority defined by the regulations. Detailed qualifications are prescribed for all drilling personnel, including service company personnel.

The requirements for the internal control plan, on the other hand, are not specified in detail. For example:

- The drilling rig shall be inspected at "short intervals."
- BOP equipment shall be pressure tested and operated at "regular intervals."
- There shall be a "fire fighting plan."
- Licensee shall ensure that the platform and equipment are in "proper working condition at all times." "Necessary" safety system shall be installed and maintained in accordance with "good oilfield practice."

However, because of the critical nature of the internal control plans in terms of their impact on operational safety, maintenance, and inspection, they are thoroughly reviewed by the NPD.

Many operators hire "classification societies" (especially Det norske Veritas) to help them develop detailed internal control plans acceptable to NPD. However, it is the licensee's responsibility under threat of criminal penalty to ensure that the plan is implemented and followed. The amount of effort expended by NPD in auditing licensees was not determined by the committee.

GLOSSARY

ANSI	American National Standards Institute
API	American Petroleum Institute
ASME	American Society of Mechanical Engineers
BAST	Best Available and Safest Technology
BLS	Bureau of Labor Statistics
BOM	Bureau of Mines
C.F.R.	Code of Federal Regulations
COE	United States Army Corps of Engineers
EPA	Environmental Protection Agency
EPRI	Electric Power Research Institute
FMEA	Failure Modes and Effects Analysis
HA	Hazard Analysis
INC	Incident of Non-Compliance
INPO	Institute for Nuclear Power Operations
MMS	Minerals Management Service
MODU	Mobil Offshore Drilling Unit
NPD	Norwegian Petroleum Directorate
NRC	Nuclear Regulatory Commission
OCS	Outer Continental Shelf
OCSLA	Outer Continental Shelf Lands Act
OOC	API's Offshore Operators Committee
OSAPE	API's Offshore Safety and Anti-Pollution Equipment Committee
OSHA	Occupational Safety and Health Administration
PINC	Potential Incident of Non-Compliance
PRA	Probabilistic Risk Analysis
SAFE	Safety Analysis Function Evaluation Chart
SPPE	Safety and Pollution Prevention Equipment
USCG	United States Coast Guard
USGS	United States Geological Survey